PRACTICAL IDEAS FOR RADIO AMATEURS

PRACTICAL IDEAS
FOR
RADIO AMATEURS

Ian Poole G3YWX

ARGUS BOOKS

Argus Books Limited
Wolsey House
Wolsey Road
Hemel Hempstead
Hertfordshire HP2 4SS

First published by Argus Books 1988

© Ian Poole 1988

ISBN 0 85242 917 7

Photosetting by Tradeset Photosetting, Welwyn Garden City.
Printed and bound in England by LR Printing Services Ltd,
Manor Royal, Crawley, West Sussex, RH10 2QN, England.

CONTENTS

ACKNOWLEDGEMENTS

I would like to thank Dave Bradshaw of Argus Specialist Publications for allowing ideas to be used in this book that I have previously included in my series 'Practicalities' for *Ham Radio Today*. I would also like to thank Mrs Joanne Rees for typing the manuscript of the book and deciphering my writing. Last, but not least, my family should not be forgotten; Pam, Stephen and Angela for their understanding whilst the book was being written.

INTRODUCTION

Amateur radio is a fascinating hobby. It covers a wide spectrum of interests from the more technical side to the enjoyment of operating. Some people prefer operating; the achievements involved in contacting new countries; the excitement of contests; or the enjoyment of talking to friends on the other side of town, or the other side of the world. To other people their interest will lie in experimenting with different types of aerial – the fascination of trying to get an extra dB or so out of the antenna. Others will enjoy constructing equipment. This may involve building new equipment from scratch. Alternatively it could involve modifying older units to improve their performance, or enabling them to be used on the amateur bands. All of these aspects of the hobby have their place.

It is probably true to say these days that most people will spend time with many aspects of the hobby. The average amateur, if he exists, will probably have some form of factory built transceiver, but will want to build various ancillary units for use around the shack. It is becoming less of an attractive proposition to consider building all of the station equipment. This is because a modern SSB or FM transceiver is so complicated that only a few people can contemplate building one, and so reasonably priced that it is not worth it.

Despite the fact that most people will buy the larger and more complicated items for their stations, there is still a lot of room for experimentation and construction. The average radio amateur still needs to have a good understanding of the equipment he installs and uses, together with the ability to construct some of the pieces of equipment around the shack. It is with this aspect of the hobby in mind that this book has been written. It contains ideas which I hope should be useful to as wide a range of people as possible.

1 SHACK IDEAS

One decision which faces every active radio amateur or short wave listener is where to set up a shack and how best to plan it. Time and thought put into planning the station before it is actually set up pays dividends later by making it easier to use.

It is possible to set up a shack in many different places and since I first became interested in amateur radio my shack has been in a variety of places. One example was a small VHF station housed in a chest of drawers. The equipment was normally hidden from view but, by lowering the front of one of the drawers which was hinged, the equipment could be operated. Another station was located in a walk-in cupboard. This was ideal in many respects because it could easily be shut off from the rest of the house. The main disadvantage was that it was difficult to get feeders in and out without having them running all round the house. Another shack was located in a loft. This was very spacious but suffered from extremes of temperature in the summer and winter. Other people have located their shacks in garden sheds, backs of garages and a whole host of other ingenious places.

Shack Requirements
Everyone will have their own ideas about how they want their shack to be set out. Some people will want to use it primarily for operating, others will want the space for construction, and some will want the shack to cater for both. Therefore, what is required will vary according to an individual's interests within the hobby. This makes hard and fast rules about the ideal shack difficult to lay down. The only principle which can be applied to every shack is that it pays to put in some thought and planning before actually setting it up.

During these initial stages of planning there are a number of points which should be considered. Features to be investigated should include the amount of room required, availability of power and access for feeders. Another point which may be of importance is the possibility of shutting the shack off from the rest of the house. If radio equipment is located in the living area it will certainly lead to comments from the rest of the family about the noise. On top of this, it is not wise to leave the equipment open to others, particularly if there are children around, because their fingers seem to get everywhere!

Even if it is possible to shut the shack off from the rest of the house it is still a good idea to have it relatively accessible. This is because it is often nice to spend just a few minutes in there to see if there is any sporadic E on 2 metres or find out if 15 metres is open. If the shack involves much effort in getting to it for one reason or another, there is a tendency not to bother to use it and to miss the sporadic E or rare DX.

Another factor to consider is the possibility of getting cables in and out. Although coax lines can be run around the house without too much difficulty they can look rather unsightly.

In view of this the shack should have ample access to the outside world. This becomes even more important if open wire feeders are to be used at any time. They certainly do not appreciate being run around the house.

Size is another important feature. In most cases the actual operating space is taken up with components, surplus equipment which may (or may not) come in useful one day, books and all the rest of the paraphernalia which goes with every amateur station. This means that the shack has to be large enough to accommodate not only the equipment itself but also the rest of the trappings, as well as leaving some space to actually get into the shack.

What's Available

After deciding upon the requirements for the shack the next stage is to look at what is available. Unfortunately the various places in which a shack could be set up seldom live up to what one would like. This is where a little ingenuity comes in useful to make the best out of what is available.

Obviously the ideal situation would be to take over a spare room and have ample space for the operating table, a constructional area, cupboards and bookshelves, but very few people are in this situation. Therefore places like large cupboards, lofts, sheds, backs of garages and so forth have to be looked at. Generally it is one's imagination and ingenuity which provides the limit to what can be done and what can be achieved.

A Cupboard Shack

One possibility for siting a shack could be a large cupboard. At first sight this might not seem to be a particularly good solution because of the space limitations. However, if it is well planned and makes the best use of all the space which is available, then it can prove to be quite an

Figure 1.1 Today's modern transceivers are very compact and enable stations to be set up in many places.

acceptable solution. In fact, one shack which I used for a couple of years was located in a walk-in type cupboard slightly less than three feet square.

The conversions to it were started by constructing a table top. This was constructed out of block board and covered with formica. The table top was then battened to the wall at either side leaving a gap at the back to pass cables down. This gap should be large enough not only for the cables, but also their associated connectors, and in particular mains plugs. If the gap is not made large enough it can be a tiresome task to take the plugs and sockets off each time a cable has to be removed.

At the front of the table top sufficient space was left for a chair when the door was closed. This meant that a chair could be kept in the shack all the time and one did not have to be filched from somewhere else every time the equipment was to be used.

In order to make maximum use of the available space plenty of shelves were put up to accommodate all the books and other paraphernalia. Extra shelves could be placed under the table as well, but they would have to be positioned very carefully so they do not interfere with the leg space.

The cupboard need not be as large as a walk-in one. Almost anything can be made to house the equipment. However, if it is small, a work top which hinges down may have to be used. If space saving ideas like this are used then it is surprising how small a cupboard can be used.

Shed Shack

One popular place for siting a shack is in an old garden shed, which can be convenient because it keeps the equipment out of the house and effectively forms another room. With a little conversion work the shack can be made to be quite comfortable and easy to use.

The first stage in the conversion is thoroughly to clean out the shed and to do any repair work which might be necessary. It is particularly important at this stage to ensure that the roof does not leak. This might be disastrous later on with expensive electronic equipment installed.

Having completed this, the mains wiring can be installed. At this stage it is worth putting in provision for more than enough sockets so that additions do not have to be made later.

Then the inside of the shed can be lined with hardboard to help keep the heat in. This can be very important in winter when there are frosts around or there is snow on the ground. The hardboard should be fastened to the structure of the shed at intervals of a few inches. It should also be cut to size so the edges can be fixed to the shed structure. This has to be done to minimise the amount by which the hardboard can warp or sag. Of course if money is not short other materials can be used to reduce this.

The next stage is to install the operating table. It is probably most convenient to construct the table *in situ* and fix it to the sides of the shed. In this way it can be made to exactly the required size so that the best use is made of the available space. As with the cupboard shack, space should be left between the back of the operating table and the wall to allow cables to be passed up and down.

With some time and effort a shed shack can be made to be very comfortable. If the walls are lined and painted, the floor carpeted and attention paid to the finishing touches it can be the ideal place to spend an evening working the DX or constructing a project.

The main drawback of a shed shack is the security. With the equipment being left unattended for long periods of time it can become a target for thieves. There are many steps which can be taken to reduce this. One can black out the window so that the equipment inside cannot be seen. Another possibility is to install some form of burglar alarm. They can be fairly easy to make, and this could be the first project for the new shack.

Passing the Cable into the Shack

One difficulty which can arise is that of devising a suitable method of passing the aerial feeders through from the outside to the inside of the shack. If care is not taken this can become a source of draughts or water leaks. This can be overcome by employing a section of one and a quarter inch plastic waste pipe to act as conduit through the wall, as shown in Figure 1.2. The outside end should be terminated in a 90 degree bend pointing downwards to prevent water entering. It is also wise to place some sealant around the point where the pipe passes into the wall to prevent any water entering the hole. Finally, the pipe can be plugged from the inside using cotton wool or something similar to prevent draughts.

This method stops the cable being chafed as it passes through the wall. It also enables new cables to be installed with a minimum amount of work. Although this method is ideal for shed type shacks it can also be used in many other applications where feeders or wires have to be passed through some form of wall.

Garage Shack

Another option is to use the back of a garage for the shack. Although not as comfortable or plush as other places, there is usually sufficient room, even with the car in, for a bench at the back of the garage, and then with the car out there will be plenty more space. There is also plenty of wall space to accommodate cabinets or shelves to hold all the components, books and pieces of equipment.

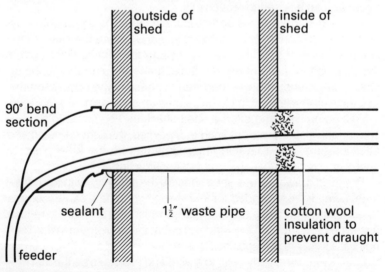

Figure 1.2 Method of getting feeders into the shed

Another drawback to the use of a garage is that it will get very cold and draughty in winter. This is bound to reduce the amount it can be used on cold days.

Loft Shack

For many people one of the most logical and convenient places to set up their shack will be in the loft. Having the shack there has several advantages. For example, it can often be reasonably easy to convert it into a shack and it will usually be fairly spacious, giving plenty of storage room. Another advantage is that the shack will be separate from the rest of the house, whilst the access to it is quite easy.

However, there are a few disadvantages which should be considered before making the final decision. Probably the first and most obvious question to ask is whether or not the loft timbers will stand the weight of all the equipment and people who may go up there. If there is *any* doubt over this it is worth consulting a friendly builder or surveyor for his opinion. The cost will also have to be considered. A floor will have to be put down, a loft ladder installed as well as having to install power and any other work which may be needed. Also remember that a loft will suffer from large variations in temperature. In summer it will become very hot and in winter it can be very cold.

Despite these disadvantages a loft can still be the most convenient place for the shack.

Spare Room Shack

Unfortunately, it is only a small number of people who possess a spare room which can be devoted to amateur radio. However, if one can be used then it can be a very convenient option to choose. Not only will it be warmer and more comfortable but it will be much easier to spend the odd five minutes in there as it will not necessitate a trip down the garden to the shed, going out into the cold garage or whatever. On top of all this, it is likely to require less work in getting the shack set up. There will be no problems of having to fit a floor as in a loft, or line the walls as in a shed. Also, there should be sufficient power already installed which means that there may be no need to alter the mains supply.

When all things are considered, using a spare room to house the shack is likely to produce the most comfortable and convenient solution with the least amount of effort.

Making an Operating Table

Wherever the shack is located, whether in a loft, spare room, shed or anywhere else, some form of table will be required. One solution is to scout around the second-hand furniture or surplus office furniture shops. By doing this, it may be possible to come by a sturdy desk or table for a fair price. Unfortunately it is not always possible to buy something suitable for the right price.

The alternative is to build one. This may seem like a daunting task to someone like myself, whose woodwork is nothing to boast about. Yet, with some thought it can be easily kept within one's capabilities. This can be done by making the design simple and only using basic woodworking techniques.

The table which I constructed used a top made from medium density chipboard, although blockboard or plywood could be used. The length will probably depend on the space available and the equipment to be placed upon it. However, the depth should be sufficient to provide about two or three inches behind the equipment for wires and connectors and then fifteen inches or so in front of it. This gives space for construction, log books, microphones, morse keys and everything else.

The table top was supported by a framework of 2″ x 1″ wood to give extra support and prevent any tendency to sag. This can sometimes be a problem, especially if heavy equipment remains on the table for a long period of time. This framework can be fixed to the table top

front and back sections
to go the full length of
the frame to hide the join   2″ × 1″ sections

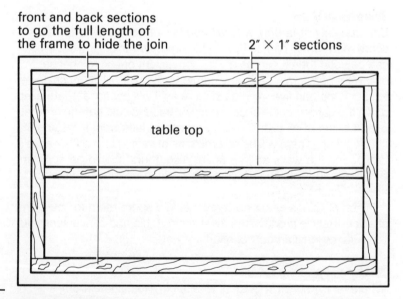

table top

Figure 1.3 Underside view of the table

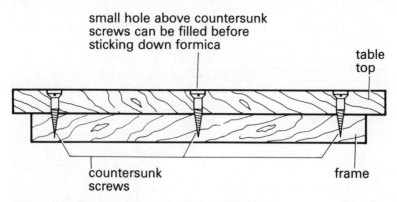

small hole above countersunk screws can be filled before sticking down formica

table top

countersunk screws

frame

Figure 1.4 Method of mounting frame onto the table top

using countersunk screws as shown in Figure 1.4. Although this does leave holes in the work top they can be filled and then the whole sur-face can be covered with formica.

In addition to providing extra support for the table top the framework enabled the legs to be attached more easily. I was fortunate enough to pick up some legs from a scrap yard. These legs had probably been used on an office desk and with one set at either end they produced a very neat and cheap solution. If you cannot get hold of anything like them, then use wood, dexion or something similar to give a firm support.

Mains Distribution

When all the equipment has been assembled on the table, one prob-lem which soon manifests itself is how to connect all the equipment to the mains. All too easily the wiring can become a messy tangle of mains leads. This can be overcome by using one or more of the mains distribution blocks containing four or five outlets in a straight line. These can be fastened to a suitable place near the back of the table, either on top or below the work surface. Then each piece of equip-ment can be permanently plugged in and turned on from its front panel switch as required.

In addition to this, the cables supplying these distribution blocks can be taken back to a common switch or circuit breaker so that the whole station can be turned off quickly and easily.

Lighting

Lighting is an important feature in any shack especially if any con-struction work is envisaged. One way of improving the lighting is to

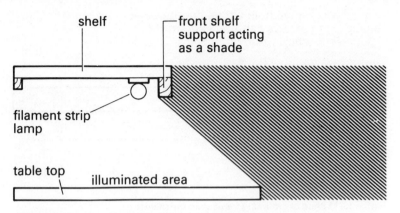

Figure 1.5 One method of illuminating the table top

install a small filament strip lamp under a shelf above the table top to illuminate the work area. The front support on the shelf can then be used to shade the lamp from direct view as shown in Figure 1.5. This approach has worked well on a couple of occasions when I have used it. However, it is worth noting that fluorescent lights should be avoided if possible because they produce a considerable amount of RF interference which extends well up even into the microwave region. If further lighting is required after this an angle poise type lamp can be used. This is ideal as it enables a large amount of light to be concentrated on the required area.

Equipment Layout
Whilst ideas for the design of the shack and table are being formulated it is worth giving some thought to the layout of the equipment so that the whole station comes together properly. Determining the best position for each unit will make using the station easier and more enjoyable. It will also ensure more efficient operating which is particularly important if any contests are to be entered.

The main transceiver or receiver should be placed in the middle of the table so that the dial can easily be seen. The tuning control should be a few inches above the table top. This enables the equipment to be tuned whilst keeping most of one's arm on the table to reduce arm ache!

If a separate transmitter is used it should be placed to the left of the receiver. Then the microphone can be held in the left hand leaving the right hand free for tuning the receiver, filling in the log book or making notes. If the station does not use a separate transmitter then this space could be reserved for a linear or possibly a second receiver.

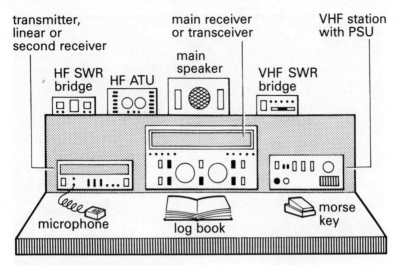

transmitter, linear or second receiver

main receiver or transceiver

VHF station with PSU

HF SWR bridge HF ATU main speaker VHF SWR bridge

microphone log book morse key

Figure 1.6 A possible station layout

If a morse key is to be used this should be placed on the right hand side of the table with sufficient table top to rest one's arm whilst sending. The lead should be long enough to be routed neatly round the equipment and log book so that it does not get in the way. This leaves space on the right hand side of the table where a small VHF setup could be placed.

It is also convenient to have a shelf above the main equipment. This can be used for smaller units like aerial tuning units, loudspeakers and SWR bridges. Keep the SWR bridge in easy view to monitor the reflected power during transmissions. Then if there is an aerial fault it will be spotted very quickly. These positions will obviously depend upon the exact equipment in the station. They have also been devised for a right handed person. Someone who is left handed would obviously have to alter them to suit. However, any station should be set up so that the pieces of equipment which are used most often are the nearest and the easiest to reach.

Decorating the Shack

Once the equipment in the shack has been set up the walls can be decorated. Maps like great circle maps, QRA locator maps, or prefix maps are very useful and give a lot of visual information very quickly. They can be easily fastened to the wall using either drawing pins or Blu-tack.

It is also nice to put up some QSL cards, particularly those showing the best DX which has been worked. Unfortunately mounting them on

Figure 1.7 A typical station set-up

the wall can sometimes damage them. Pinning them to the wall obviously puts a hole in them, and Blu-tack can leave a mark after a while.

To overcome this problem it is possible to mount the cards onto a postcard or some other suitable card first using photograph corners. Then the postcard can be pinned or stuck to the wall, leaving the QSL cards free from damage.

Postcards and photo corners are often found around the home, or they can be bought quite cheaply from a local stationer. This makes it a cheap and easy way of displaying QSL cards around the shack without spoiling them.

General Safety

It is probably true to say that safety standards in shacks have improved over the past few years. This is partly as a result of an increased awareness of the hazards, and partly due to the voltages in equipment being lower. Another reason is that more commercially-made equipment is being bought. This has to comply with certain safety standards before it can be imported or sold. However there are still a number of safety precautions which can easily be incorporated into the shack.

The first is to ensure that all equipment is properly earthed. Sometimes there is a tendency to leave the earth off some pieces of equipment but this can be dangerous as it means that the whole of the case can rise to mains potential under certain fault conditions.

Another useful idea is to fit a master switch to turn the mains off from the shack or operating table. This should be very accessible so that, if something does go wrong, all the mains can be turned off quite easily. It is also a good idea to let the rest of the family know where it is.

An earth leakage circuit breaker, or better still a residual current circuit breaker, is worth putting in. The earth leakage circuit breaker trips the mains off if current over a few tens of milliamps flows along the earth line. So if there is any leakage to the earth line and the possibility of metalwork becoming live, the circuit breaker will turn the mains off. The residual current circuit breaker is more useful. This trips out if there is an imbalance between the current flowing in the live and neutral lines. This is better as it is more likely to prevent serious electrical shocks or leakage along paths other than the earth wire.

Whilst on the subject of electric shocks, it is worth mentioning that all equipment carrying hazardous voltages or high levels of RF should be enclosed in cabinets. This is particularly important if visitors are likely to come into the shack at any time.

Finally, radiated RF should be kept away from inhabited areas of the house. Although it is unlikely that any harm could be caused by low powers and non directional aerials, this may not be true where high powers and directional aerials are used. However, as it is difficult to assess field strengths at a particular place, it is best to keep all RF at a distance.

These ideas represent only a few of the ways for keeping a shack safe. There are many other points which can make it safer. A general awareness of the dangers which might arise is the best solution. Then the shack will be a safer place for you and any visitors who may call in.

2 AERIAL IDEAS

The aerial system in any amateur radio station plays a vital role. In fact the whole operation of the station depends upon its efficiency. Because of this, experiments with different types of aerial, or improvements to existing ones, are always interesting and worthwhile. They can pay dividends by increasing signal strengths and enabling contacts to be made over greater distances or under poorer conditions. So whatever one's brand of amateur radio, be it construction, DXing, local nattering or whatever, an aerial will always be required.

Aerial Lengths

There are a number of aerials which rely on the use of a half or quarter wavelength for their operation. One of the most popular must be the half wave dipole. However, the physical length of the aerial does not correspond exactly to the length of a half wave in free space. There are several reasons for this. One is the detuning effect of whatever is holding the aerial at either end, together with other stray effects. Another is the diameter of the wire. Fortunately, the factor by which the aerial is shortened is generally about 5% at HF. As a result it is possible to calculate the length of a half wave aerial using the formula:

$$I = \frac{470}{f}$$

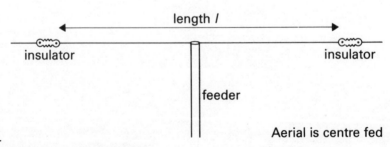

Figure 2.1 A half wave dipole

In this case f is the frequency in MegaHertz for which the aerial is to be used, and I is its length in feet.

As the formula is only a generalisation and the length of the aerial will vary slightly from case to case, it is usual practice to make the aerial a bit longer than the calculations would suggest. Then it can be 'pruned' to give the best results on a particular portion of the band.

Mounting Wire Aerials to Trees

When putting up aerial systems it is not always necessary to erect a mast. It is often quite satisfactory to fix it to a convenient point on the house or even a suitably placed tree. In the event of a tree being used some account will have to be taken of the wind movement otherwise the aerial will not last very long!

In order to get round this problem, a system is needed which will keep the tension constant whilst allowing the tree to move freely in the wind. Figure 2.2 shows a widely used system which is very success-ful, but which only requires the use of a pulley and weight. By using these extra components a constant tension, determined by the weight, is maintained but any movement in the tree is allowed for by the pulley.

When choosing the size of the weight and hence the tension, care should be taken not to make the tension too large. It should be suffi-cient to take up most of the sag. However, it should be remembered that some sag will always be present and to reduce it beyond a certain point will require a large increase in tension.

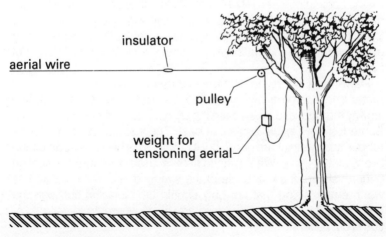

Figure 2.2 Method of fixing an aerial to a tree

Once the tension has been set up the system should be checked to ensure it is free to move and is not likely to catch any branches in the wind.

Running Coax Underground

In many amateur radio stations it is necessary to take a coaxial feeder across a lawn or a section of garden to the aerial. There are two ways in which this can be done. Either it can be taken above ground which can be unsightly, or it can go underground. When running coax underground it can be conveniently run through some hosepipe. This serves two purposes. Firstly, it provides some protection to the feeder and secondly, if it is ever necessary to change the feeder it is simply a matter of pulling the old one out and using it to pull the new one through.

The major precaution to be aware of when using hose is to prevent water getting in. Obviously if water does manage to get in and there is no way out it will just stay there. Ideally the hosepipe should be on a slope with the bottom end open to the air to allow any water which does get in to escape. Then the top end can be sealed to prevent any water getting in. For this any sealant such as the silicon rubber bath sealant is ideal.

Preventing Moisture Entering Coax

The weather and damp in particular are the perpetual enemies of any aerial installation. Coaxial feeder is often prone to suffer from water entering it and causing the loss in it to rise dramatically. It is very easy for the coax to collect a lot of water. Take, for example, a simple wire dipole. The water will tend to collect on the wire and run towards the lowest point on the aerial, which will be the centre (if it is supported at either end). Then the water will run onto the coax or into it if it is not sealed.

However, it is easy to seal the end of the cable using a variety of materials. Tape on its own should not be used as water will easily find its way through, and in any case the tape will come away after a while. Some form of flexible sealant is best. The various types of silicone rubber bath sealant are good, but it is also possible to use something like Evostick glue. Whatever sealant is used it should be applied generously around the whole of the exposed end of the cable. It is worth emphasising that not only should the dielectric between the conductors be protected, but also any place where the outer insulation is cut back.

If the coax is protected in this way, then it will not only last far longer but it will also save the expense of replacing it too often. This can be quite a saving in view of the cost of cable today.

Assessing the Loss in a Feeder

Despite all the precautions taken to prevent the weather getting into coax, there is always the possibility that it will eventually find its way in somehow. Either the sealing at the end of the coax may fail or the outer insulation may become damaged and let in moisture. The degradation to the coax performance which results from this will be gradual and often pass unnoticed, so it is worth carrying out a periodic check on the feeder to assess its loss.

Another instance when it is worth checking the performance of some coax is when a surplus length is to be brought back into commission. With the ever increasing cost of coax it can become a necessity to re-use any serviceable lengths and not just discard it as possibly being suspect.

There are several ways of assessing the loss introduced by a length of coax. Probably the most obvious way is to terminate the feeder in a resistive load and measure the power at either end. This method is perfectly sound in theory but it is not always particularly viable in practice. Firstly, it requires access to a power meter and secondly two measurements have to be made – one at either end of the coax. Whilst it will be comparatively easy to measure the power in the shack, it may not be so easy at the remote end if it is up a tower or in some other inaccessible place.

One way which overcomes these problems to a large extent is to substitute the matched load for a complete mismatch such as a short circuit and then measure the VSWR. In this case all the power reaching the short circuit will be reflected and this can be measured by a VSWR meter. This method works because the loss in the feeder will reduce the amount of reflected power which the meter will see. Using the VSWR reading obtained it is possible to deduce the loss in the cable.

When using this method of measuring the feeder loss, a few practical points should be noted. Probably the most obvious is that the power source should be capable of operating with high values of VSWR without any risk of damage. In addition to this the power levels should be kept as low as possible because the high voltages and currents generated by the high level of standing waves may damage the feeder. It is also worth remembering that the loss in any given piece of coax will rise with frequency, and any measurements should be made at the highest frequency for which the cable will be used.

Finally, the short circuit which is applied to the cable should be as short as possible. This is particularly important for frequencies above 30 MHz or so where even short lengths of wire may possess a significant inductance which could invalidate the results.

Having obtained the reading for the VSWR it can be converted into a figure for the loss by referring to the Table in Figure 2.3. Then it is just a matter of assessing whether the loss is acceptable.

Figure 2.3 Table to Convert VSWR Reading to Feeder Loss for a Short Circuited Feeder

VSWR Reading	Feeder Loss dB
1.02:1	20
1.05:1	16
1.10:1	13
1.20:1	10
1.4:1	7.5
1.6:1	6
1.8:1	5
2.0:1	4.5
2.5:1	3.5
3:1	3
4:1	2
5:1	1.75
6:1	1.5
10:1	0.75
15:1	0.6
20:1	0.4

Aerial Wire

It is generally not realised how much ordinary copper wire can stretch if it is placed under tension. As copper is a very ductile metal, the wire made from it can elongate quite considerably when a length of it is pulled taut. For example, a length of multi-stranded PVC covered wire can extend as much as ten or fifteen percent when it is held under sufficient tension to keep the aerial from sagging. This problem is particularly acute for the lower frequency bands where longer lengths and higher tensions are required.

This wire elongation will correspondingly decrease the resonant frequency of the aerial. If the aerial is centre-fed using coax this will cause the SWR to rise over a period of time and affect any 'fine tuning' that may have been carried out. There are two solutions to the problem. The first is to use only sufficient tension on the aerial to reduce sag to an acceptable level. The second is to use a large enough gauge of hard-drawn copper wire which will elongate much less than other varieties of copper wire.

Longwire Pros and Cons

The longwire is a very convenient aerial to use but it is not always the best. It is widely used because it is versatile and easy to construct. All that is needed is a length of wire, which may or may not be cut to a resonant length, a few insulators and an ATU. It is also easy to erect and it does not need an expensive feeder. Then it can be made to look almost invisible.

In many ways the longwire (or end fed wire as it should more correctly be called) has a lot of advantages. However, it does have a few drawbacks, particularly for the transmitting amateur. Firstly the aerial has no feeder and it starts radiating as soon as the wire leaves the ATU. This means that some of the wire required for radiation will be absorbed by nearby objects. It can also mean that there are fairly high power densities in the shack. This can be bad from the point of view of health for anyone in the shack if high power levels are used. Whilst it is not easy for the radio amateur to measure power density levels, it is worth bearing in mind the comments which have appeared in the amateur press, and keep these levels to a minimum.

Another problem which often manifests itself when longwires are used is TVI. There are two main reasons for this. The first is that the aerial radiates as soon as it leaves the ATU. If the shack is in or near the house, then it means that the source of the RF is much nearer the television itself. This will make TVI more likely to arise. In fact siting an aerial, whatever type it may be, away from the house can often overcome TVI problems.

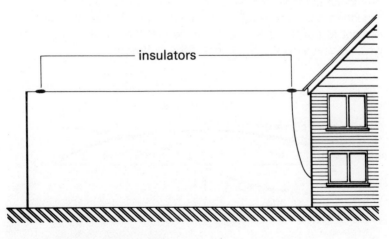

insulators

Figure 2.4 A typical longwire aerial

The second way in which longwires cause interference is by RF being conducted along the mains. As the longwire is an unbalanced aerial which needs tuning against an earth or counterpoise, it often happens that the mains wiring becomes part of the system and the RF is carried everywhere.

Bending Aerials

A dipole for one of the HF bands is not usually thought of as a large aerial. But in today's postage-stamp size gardens it can be difficult to put up even a modest sized aerial. Fortunately, it is possible to make many types of aerial fit by bending them and usually this does not affect their operation too much. The reason for this can be seen by looking at the current and voltage distribution for an aerial, and in particular a half wave dipole as shown in Figure 2.5. From this it can be seen that the current is a maximum at the centre, falling to zero at either end. As it is the current flowing in the aerial wire which actually causes the radiation the centre is where the majority of the radiation occurs. Therefore, if the centre is left undisturbed, the aerial performance will not be affected too much. The main precaution is to ensure that the ends of the dipole do not pass too close to any objects which are likely to detune the aerial.

It is obviously possible to bend a dipole in many ways. Two possibilities are shown in Figure 2.6. In the first instance the ends of the dipole are bent downwards, and in the second the ends are bent to either side.

Another popular way in which the trusty old dipole can be bent to give the inverted V is shown in Figure 2.7. By using this configuration the space taken by the aerial is reduced. In addition to this, it only requires the use of one tall mast or other support, making it more convenient to put up. It is also fortunate that the centre of the aerial, which radiates most of the signal, is the highest point.

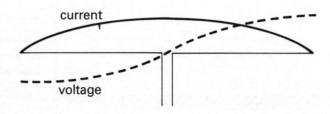

Figure 2.5 Current and voltage distribution for a half wave dipole

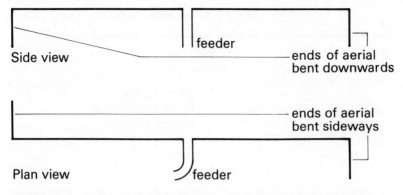

Figure 2.6 Two ways of bending a dipole without significantly degrading its performance

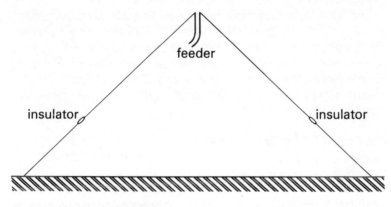

Figure 2.7 The inverted V dipole

A Cheap Multiband HF Antenna

The drawback with a dipole fed with coax is that it can only operate over a single band or bands of frequencies. Although this may be quite satisfactory if operation is to be restricted to one or two bands, this is not usually the case. One way of overcoming this is to use some form of trapped aerial. This may not always be satisfactory either, particularly if a large number of bands are to be used.

One very attractive way of overcoming the problem is to use a doublet fed with open wire feeder, as shown in Figure 2.8. Whilst it may seem old fashioned, it has the advantage that it has been proved to work by many people over the years. It is also a truly wide band aerial which can operate at frequencies above the point where it is a half wave dipole.

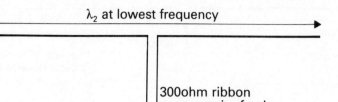

Figure 2.8 The multiband antenna

When using an aerial of this type an ATU must be used. It should also be one able to match the standard unbalanced 50 ohm line from the transmitter to a balanced open wire or ribbon cable.

The feeder is not as difficult to come by as one might expect. It can either be commercially-made 300 ohm ribbon cable or it can be home-made quite easily. Ideally the feeder should not be excessively long as the aerial system will then have a narrow bandwidth and will require retuning when the frequency is changed. On top of this the feeder should be kept reasonably well clear of nearby objects which may unbalance it, causing losses.

The Correct Type of Ribbon Feeder

When putting up aerials using the commercially-made 300 ohm ribbon cable it is very important to use the correct type. The cheap translucent feeder which is available from many component stockists is adequate for indoor use only. If it is used outside, the plastic absorbs water causing its impedance to change and the loss to rise. Interestingly, the feeder soon dries out in the sun and can continue to be used but, if the cable is left outside for any length of time, the absorbed water causes the internal conductors to 'weather' and the cable as a whole to be degraded.

The best cable to use is the black plastic ribbon which has oval shaped holes in the dielectric. This is a bit more expensive but it is far superior and quite suitable for outside use.

Open Wire Feeders

Even though 300 ohm ribbon cable is easy to use and readily obtainable, it should not be forgotten that it is possible to make open wire feeder. In fact it is surprisingly easy to make, and very tolerant to variation of some of the parameters such as spacing, wire gauge and so forth. In other words, it can be made up from any suitable odds and ends which are available.

All that is needed is the wire and the spacers. The wire should be thick enough so that its resistance along the total run remains low. Beyond that, it can be hard drawn, PVC covered or almost anything. The spacers too can utilise what is available. Suitable pieces of plastic could be drilled with two holes, one for each wire in the feeder. Failing the homebrew approach, it is possible to buy them.

The wire spacing is not particularly critical. About four inches is quite suitable, and this will require spacers to be placed every two or three feet. It is possible to place them closer together, but this will tend to increase the losses during wet weather.

Earth Systems for HF Verticals

Trapped verticals can make a neat and effective aerial for multiband HF operation. They are compact and can even fit into a postage-stamp sized garden if they are ground-mounted and tuned against earth.

If this approach is adopted, it is absolutely necessary to ensure that a good earth is used. It should obviously be directly under the aerial itself and should also have a low DC resistance if the efficiency of the aerial is not to be degraded. The earth resistance will be dependent on two factors. The first is the amount of metal which is buried – the more buried metal the better. It is even possible to lay two radials to reduce the RF impedance. The second factor is the local earth conductivity and unfortunately there is not a lot which can be done to improve it, apart from moving to a different area! Broadly speaking areas of good fertile soil are likely to have good conductivity but areas of dry, sandy soil, or where there is only a little top soil above the bedrock, are likely to have poor conductivity.

So if you live in an area of poor earth conductivity, a ground mounted vertical may not perform to its best and a true ground plane could be better. However, if the ground conductivity is good then the aerial should perform well.

A Cheap VHF Vertical

There are still a number of standoff type insulators available at rallies and club junk sales. Although they may not be needed in exactly their original role they can often be put to good use with a little imagination.

One application can be as part of a VHF vertical. The insulator shown in Figure 2.9 can be used as the base and an old car aerial used as the vertical element.

These insulators usually come complete with a screw and wing nut. However, the problem still remains of how to connect the coax inner to the car aerial itself and how to fix it mechanically. This can be done

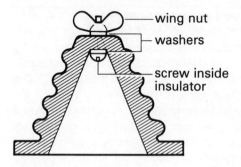

Figure 2.9 A standoff insulator

by making a small adaptor out of brass on a lathe. If one has access to a lathe this should present no problem but, for those who do not possess one, a friend or fellow radio amateur who has a lathe may be able to come to the rescue. The adaptor should be made to take the screw from the insulator at one end and the aerial at the other.

When the adaptor is complete the section of car aerial should be soldered in, so that a good connection is made. Then the adaptor can be fixed to the insulator using the screw. The centre conductor of the coax can then be connected to the screw either by using a suitable solder tag, or simply by using the screw as a screw terminal.

The outer braid from the coax should be taken to some radials and the length of the car aerial altered to give a suitable match.

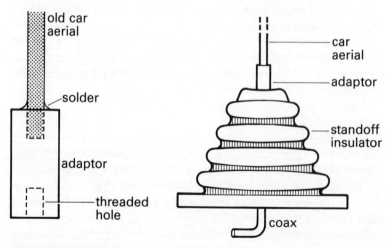

Figure 2.10 Adaptor to join car aerial to insulator

Figure 2.11 The finished aerial

3   CONSTRUCTIONAL IDEAS

Construction is at the heart of amateur radio. In the early days of the hobby it was essential for people to build most of their equipment but things have totally changed and there are many good reasons for buying ready-built equipment, either new or second-hand. It is often cheaper to buy than to build, it saves a lot of valuable time and also the finish on commercially-made equipment is usually better. In spite of all this there is still a large place in the hobby for construction. Many people who own a good factory-built set often still want to enjoy the challenge and enjoyment of building some of their own equipment.

So, all in all, there are still a lot of people who enjoy constructing part of their equipment and consider it an essential part of the hobby.

Mechanical Construction

The mechanical side of almost any project is the part which receives least attention, because it is usually felt to be the least interesting. As a result it is quite usual to see some quite complicated or well-made boards just sitting on the bench or mounted in a makeshift box. Unfortunately, this can lead to things becoming damaged, or other problems like RF pickup caused by the lack of screening.

However, it is possible to put many of these projects into quite presentable cases with a minimum of effort. Once this is done it will sometimes enable the circuit to work better, or at least it will give it some protection and prevent it from becoming damaged.

Using an Abrafile
One tool which comes in handy when doing metalwork is an Abrafile. It is possibly not well known to many home constructors, but it is extremely useful for cutting large or irregular holes in sheet metal. It is ideal for making cutouts in panels when the correct size of drill is not available or if the hole is not circular.

Basically, an Abrafile is a long circular file about a sixteenth of an inch in diameter and is designed to fit a hacksaw frame. If a hacksaw is already available, all one needs to buy are the Abrafile itself and the small fitments for either end of the file. All of these are readily available from hardware shops.

To use the Abrafile, the first job is to drill a small hole large enough to take the file. The file can then be passed through the hole and attached to the hacksaw frame, and should be kept under tension by tightening the nut at the end of the hacksaw. It should be given sufficient tension to prevent the file bending unduly, but not so much that it snaps.

Having set it up, it is then a simple matter to make the cutout by following an outline which should have been marked previously.

Professional Finishes on Homebrew Equipment

One problem always present for the home constructor is that of achieving a reasonably professional finish to the metalwork with a minimum of equipment and time. One way of doing this is to use a ready painted case with a front panel. These cases are available from most component stockists.

The main drawback, apart from the extra expense, is that the paintwork is very easily spoiled when the hole drilling and cutting is being done. One way of overcoming this is to cover all the vulnerable areas with masking tape, remembering that any surfaces which are left uncovered will be at risk. It is better to cover too much of the case than too little. One further advantage of this is that it is possible to write or draw on the tape and in this way, the layout can be planned before the drilling starts.

If the project becomes delayed, as often happens, and the tape is left on the case for any length of time, the tape may become difficult to remove. If this does occur, then remove as much of the tape as possible. The remainder can then be cleaned off with the help of a freon-based solvent, often available at electronic component stockists.

Figure 3.1 A painted project box

Centre Punches

When starting the metalwork on almost any job, the major task is to drill all the holes. These holes need to be positioned fairly accurately so that the boards and other components which have to be bolted down actually fit. Also, the front panel will look much smarter if all the holes are in the correct places and not randomly offset from their required positions. Unfortunately when drilling holes in metal, or anything else for that matter, the drill will tend to wander and start the hole where it wants. To overcome this, a small starter 'hole' or mark should be made in the exact position of the centre of the hole to be drilled. By so doing, the drill will start the hole in the correct position and not wander.

The starter 'hole' or mark can be made using a centre punch. This can be a purpose-made centre punch but, if one of these is not available, an old nail can be pressed into service. The starter 'hole' should not be very large, just enough to keep the drill from wandering. It should make an indentation into the surface of the metal but it should not actually dent it.

Front Panel Labelling

Once the metalwork has been completed, the front panel can be labelled. This improves the appearance of the project and gives it a much better finish. Unfortunately it is a job that often gets left. One reason for this is that it does not improve the electrical performance of the unit. Another reason may be that there does not appear to be a cheap and hard-wearing method for labelling, and any attempts which are made with transfers soon rub off.

One method which has proved to be neat, hard-wearing and not too expensive is to use Letraset or another similar form of transfer. Once the labelling has been completed, it can be protected by covering it with clear Fablon or adhesive book covering. When applying the adhesive covering it is best to remove all the panel mounted components such as switches, volume controls and so forth. Once this has been done the covering can be applied. Care has to be taken whilst doing this to ensure that no air bubbles are trapped or that the covering is not creased. Finally, the edges can be trimmed, the holes in the panel cleared and the components replaced to give the desired finish.

An Idea for Finishing an Aluminium Panel

It may not always be possible to use a ready-painted case, or one which has some other form of neat finish on the front panel. When this happens the front panel is usually just bare aluminium which will quickly look grubby and untidy.

It is possible to improve the look of the panel by giving it a 'grained' effect. This can be done by brushing it with wire-wool or a Brillo pad. If this is done carefully, ensuring that all the brushing is in exactly the same direction, then it can give a pleasing finish. The labelling can then be added and the panel can be covered with clear adhesive as before. Not only will this protect the labelling, it will also keep the panel clean.

Electronic Construction

The electronic side of construction attracts more interest and attention than the mechanical side. Circuits for projects of all sorts are regularly published in the magazines and they can be built up in a number of ways. Either printed circuit boards can be used, or possibly one of the types of matrix board. Both of these methods work well but each one has its advantages and disadvantages. In addition to the choice of the type of board to be used there are also many ideas which can be built into the unit as a whole. They can make it easier to construct, improve the performance, or just make it look better.

Soldering and Tag Strips

Very often when one is delving into the depths of a piece of equipment with solder tags it is difficult to remove a wire or component. This is because the wire has been wound through and round the tag. This is often done during the construction to keep the component or wire in place prior to soldering and to make the joint very strong mechanically. As far as the amateur is concerned it is much easier to place the lead through the hole and possibly bend it slightly to make a neat job before finally soldering. If this method is adopted, then it makes the job of modifying the unit or servicing it later very much easier and quicker. The fact that the component lead has not been wrapped right around the tag will not affect the electrical performance of the joint. Mechanically it will not be as strong, but in most cases this should not be a problem as the joint should not be under tension anyway.

Track Cuts in Stripboard

Stripboard, or Veroboard, with strips is widely used for building up circuits. It is convenient because it enables projects to be built up using a minimum number of wire inter-connections, unlike the plain matrix board which requires each component to be wired to the next.

One major problem occurs when the strips are cut, which is usually done using either a special cutter or a suitably sized drill. When either of these tools are used, it is quite common for small whiskers of track to remain almost unseen but still connecting the two sections of strip together. Also, whiskers of copper from the cut can sometimes short to the adjacent track.

To overcome these problems, each cut should be carefully inspected to ensure there are no whiskers left. Then a screwdriver can be run up and down the board between the strips to ensure that there are no whiskers of copper shorting between the adjacent tracks.

Desoldering Braid

Removing components from printed circuit boards, or just desoldering them, is not always easy with the tools available to the average amateur. Some people will have solder suckers, which makes the job easier, but by no means everybody has one.

Another way of removing solder is to use desoldering braid. This is a braid made up from fairly thin strands of copper wire and, when it is placed onto a soldered joint with a soldering iron on top of it, the solder is absorbed into the braid leaving just the joint almost completely free of solder. The braid is fairly cheap and it is worth keeping some in the tool box for the occasions when it is needed.

Plain Matrix Board

Although the matrix board or Veroboard with strips is widely used it can often become rather restricting. The layout can become rather convoluted, as it has to be made to fit the tracks, and mistakes are easily made.

Figure 3.2 A project built on matrix board

Another form of board is plain and has no strips. It can be used with the corresponding pins to provide another way of building up a circuit. One advantage of it is that the circuit can be laid out in a manner which resembles the circuit diagram. If this is done it makes the job of construction easier and reduces the likelihood of mistakes. In addition to this, if all the capacitors, resistors and other similar components are mounted on pins it makes the job of changing them much easier. It also means that all the intercomponent wiring can be kept on the underside of the board resulting in a neat, compact and reliable circuit board.

Etch Resist for Printed Circuits

There are many ways in which printed circuits can be made. Many component stockists offer kits of various forms. Alternatively the etch resist transfers, photo resist, ferric chloride and all the other paraphernalia used for making PCB's can be bought separately. Most of the methods give good results but often the amateur experimenter will want to throw a circuit together quickly to try out a new idea, or he may not want to go to all the expense and trouble of making the board look quite so neat and tidy.

In fact, all that is needed on the board is something to resist the attack of the ferric chloride. Probably the quickest way of achieving this is to use an etch resist pen. This is a fibre tipped pen which uses a special ink and is both quick drying and resistant to the etching solution. If a proper etch resist pen is not available it is sometimes possible to use a spirit-based permanent marker pen. Before using one of these, it is worth experimenting with a small off-cut of copper clad board to ensure it works.

Using these pens, the required copper pattern is drawn straight onto the board. The best way to do this is to start in pencil as mistakes or changes can be rectified quite easily. Only when the pencil layout is satisfactory need the etch resist pen be used.

If an etch resist pen is not available then it is possible to use one of the various forms of cellulose paints, such as nail varnish. Even though these paints resist the action of the etch very well they do not give a very accurate or neat finish. Because of this they are best used only for large areas or if the local shop is shut and no other method is available.

Component Clearance Holes in Earth Planes

When making up printed circuit boards for RF circuits it is usual, and certainly a good idea, to use double-sided board. This enables the tracks to be run on the under side of the board, leaving the component side intact, or nearly intact, for use as an earth plane.

When doing this, the earth plane has to be cleaned from around the component holes to prevent them from being shorted to earth. In commercial companies a separate artwork is generated for the component side to do this. However, for the amateur to have to enable each clearance hole to be etched and formed in the correct place represents a lot of extra work. It is much easier to leave the earth plane intact during etching and only make the clearance holes after the component holes have been drilled. This is easily done by using a drill with a shallow angle to countersink the component holes as required. Provided the countersink is not made too deep, this method performs very well and saves all the effort of having to sort out where to etch the copper in the earth plane.

Putting Screens on PCB's

When building or designing printed circuit boards it often happens that some sort of screen is needed. This obviously adds extra work which can be inconvenient and, on top of this, it may not add to the looks of one's prized circuit board. Fortunately most RF boards are double-sided and the top side is used as an earth plane. This can be very convenient because the screens can be easily soldered to the top side of the board where most of the stray pickup occurs.

One way of making up a screen very conveniently is to use a small offcut of double-sided board. This can be cut to the correct size and neatly soldered in place as shown in Figure 3.3. Once in place it is mechanically strong as well as giving a high degree of screening.

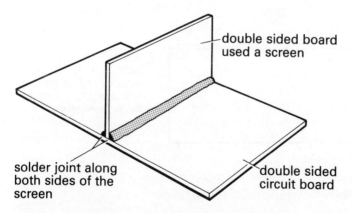

double sided board used a screen

solder joint along both sides of the screen

double sided circuit board

Figure 3.3 Double-sided copper clad board used as a screen

Keeping Ferric Chloride

Ferric chloride is not the easiest solution to keep for any length of time. It undergoes a rather complex reaction to form ferrous chloride, which is of no use when etching circuit boards. As a result, it is best to keep the chemical in its anhydrous (free from water) state and not make it up until it is required.

Once the solution has been made up it should be kept cool to slow down the rate of the reaction. In addition, if a used solution is chilled, some of the etched copper will tend to settle out as a sludge at the bottom of the container. Then the solution can be poured into another container and the sludge discarded.

PCB Heatsink

Quite often there is a need to mount a component on a small heatsink. One example might be a voltage regulator which is not dissipating enough heat to warrant a full heatsink, but needs a small one as a precaution. One solution to this problem is to use the copper on the printed circuit board itself. Obviously, when using this sort of approach there should be sufficient copper to dissipate the heat. This can be done by using double-sided board. The underside can be used for the tracks as normal but the top side can be kept intact, similar to an earth plane used on an RF board.

Although this idea may not appear to conform to the normal methods of dissipating heat there are some integrated circuits which are designed to use it. One such chip is an LM380 which uses the six pins nearest the centre, as shown in Figure 3.4, to remove the heat

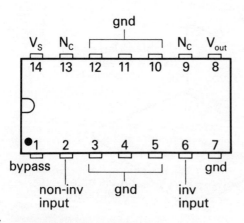

pins 3, 4, 5, 10, 11, 12 connected to ground are used to conduct heat from the ic

Figure 3.4 Pin-out of an LM380

from the IC onto the board. Sometimes a heatsink connected to these pins is used if higher powers are needed.

The main advantage of using the printed circuit board in this way is that it is very convenient and saves having to mount the component onto a separate heatsink with all the extra wiring which that may involve. It also proves to be perfectly adequate for small amounts of power, and if it works why not use it?

Using IC Holders

Most people will build up circuits of one sort or another using integrated circuits. In fact as the variety of IC's increases so does the frequency with which they are used. Using IC's is fine if the circuit works first time. However, as most circuits will have some form of bug in them, the fault has to be traced through. Then if the fault appears to be a chip it has to be removed.

Unfortunately it is not easy to remove an IC from a board because it is very easy to damage either the board or the IC. This can be particularly annoying if one finds out later that the IC was not faulty after all. As a result, it is much easier to use IC holders wherever there are the dual-in-line type IC's. It is then a simple matter to remove or change a chip and check the faults.

Possibly the major drawback when using holders is that they can be a source of faults themselves. Whilst this is perfectly true, it usually happens that once the IC has been correctly inserted there are very few problems. The major source of failures occurs when inserting the IC's into their holders. Either a pin can become bent under the IC or it may go outside the holder. In both cases these problems can be minimised by taking a little extra care when inserting the IC's. If this is done, it should make the process of checking or servicing the board very much easier.

Lacing Cord

It is easy for projects with a lot of interconnection wiring to become untidy. One way of improving the appearance of a job and making it much neater is to gather up all the cables into a loom. Whilst this may seem rather daunting it is not as difficult as it may appear at first sight.

Firstly, when wiring up a unit the route to be taken by the loom should be chosen. Then the wires can be cut to the correct length. If anything, it is wise to make them a little too long as any extra length can be lost by looping the wire from the loom onto the component.

Once the wiring is complete the loom can be laced together. It is best to use proper lacing cord which can be obtained from most component dealers. It makes a much better job than trying to improvise.

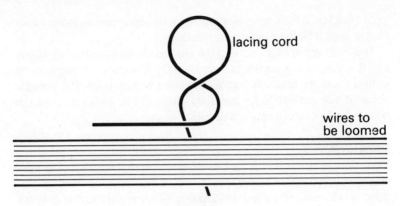

lacing cord

wires to
be loomed

1st stage – make a loop and twist

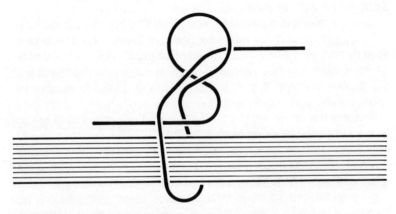

2nd stage – take lacing cord around the
loop, and back up through the loop

Figure 3.5 How to loom wires

The basic knot used is shown in Figure 3.5. It is best to practise it
first before trying it out on the unit itself. Although the diagram of the
knot may look a little complicated it is quite easy to gain the hang of it.
The knots should be tight enough to stay in position, and the cord between the knots should not have any slack. Starting and finishing the
lacing can be accomplished by tying two knots on top of one another.
This should prevent them from slipping and spoiling the loom.

Sleeving Joints

Another wiring aid which is seldom used in amateur radio projects is
sleeving. This can be used to cover exposed terminals or soldered

joints and is particularly useful where there could be exposed terminals carrying high voltages. Sleeves are also useful when wiring to small multi-pin connectors where the exposed wires could bend slightly causing shorts between the pins.

Sleeves come in various sizes and they are usually made of rubber so that they can stretch slightly when they cover a joint or terminal. This makes them stay in position once they have been fitted.

Although sleeves do not have to be used on all exposed joints they can be very useful in some instances.

Safety Precautions

One item which should feature very strongly in any piece of home-brew equipment is safety. This is particularly important when high voltages are generated in the equipment or if it is powered by the mains. In fact, whatever the equipment, precautions should be taken to ensure that everything is as safe as possible. This should prevent not only oneself, but also friends or family from getting a rather unwelcome and unpleasant shock.

The most important precaution is to ensure that all metalwork is securely grounded to prevent anything from becoming live if there is a short circuit. Another precaution is to protect any live terminals. This

Figure 3.6 A home-made power supply – note the use of sleeves and lacing cord

can usually be accomplished using sleeves as already mentioned. In addition, the cables connecting the equipment to the mains should be carefully wired up. Often one sees mains cables which are wired up carelessly; maybe the earth connection has been left off or the cable grip has not been tightened.

Although these are a few examples of ways of improving safety it is best to develop an awareness of possible hazards, and then to apply common sense to overcome them. Unfortunately this does require a certain amount of extra time and effort. However, when considering some of the potential hazards which there are in the shack is not the extra time worth it?

Removing Kinks from Lengths of Wire

In almost every radio amateur's shack there is one of the proverbial junk boxes, an Aladdin's Cave filled with all manner of bits and pieces. Caught up amongst everything else, there are no doubt odd lengths of wire thrown in to be used at a later date. This wire quickly becomes full of kinks and bends which can be difficult to remove if the right tricks of the trade are not used.

One method for removing these kinks is quite simple. The wire is wound once round a screwdriver shaft, or other implement. Having done this the wire is simply pulled so that the entire length passes round the shaft as shown in Figure 3.7. By doing this the wire loses all its kinks or bends and it can be re-used.

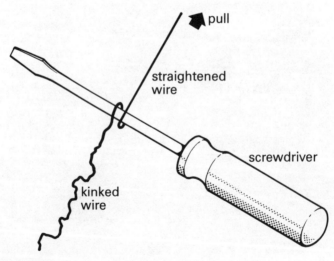

Figure 3.7 Straightening out kinked and bent wire

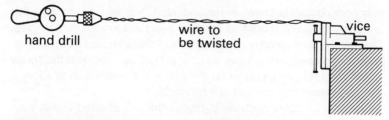

hand drill   wire to
be twisted   vice

Figure 3.8 Method of obtaining neatly twisted wires

Twisted Pairs of Wire

Sometimes it is useful to have two wires twisted together. This helps to keep the wires together, making a job neater and also reduces pickup as both the wires are subject to the same fields.

The easiest way to do this is to cut two wires of the same length, preferably choosing two different colours to enable each wire to be identified later. One end of each wire should be placed in a vice, or something similar, and the other ends placed in the bit of a hand drill. Once they are secured the hand drill is turned to put slightly more than the required amount of twist into the wires. The extra twists are put in because once the wires are relaxed they will untwist slightly.

The advantage of this method is that it gives an even twist to the wire, making a much neater job, as well as being easier than other methods.

RF Screening and Painted Cases

The ready-painted cases which have been mentioned previously are ideal in many respects but there are a few points to watch when using them. Many amateur radio projects require careful screening and this may not be provided by these cases if some precautions are not taken.

Some of these cases consist of two metal sections, as shown in Figure 3.9. As each section is painted it means that at best there is

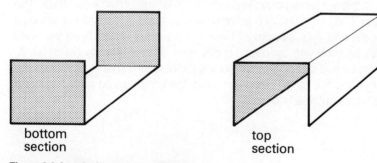

bottom
section   top
section

Figure 3.9 A typical two piece project case

only poor contact between the lid and main chassis. To overcome this problem, the paint should be scraped away around the screw holes. This can be done neatly so that it does not show and spoil the appearance of the case. Then a star washer should be used with the screw to bite through any oxide when the screw is tightened. In this way a good solid electrical contact can be made.

Not only adjacent sections of metalwork are affected in this way. Many connectors, such as BNC or SO239, rely on an earth connection to the chassis. Again, the paint should be scraped away from around the holes. It may seem a shame to spoil a nicely-painted case in this way but it can be done neatly on the inside of the box and only around the connector itself so that it cannot be seen. If it is done like this, then it should not spoil the appearance of the case but improve the electrical performance of the unit.

Other Ways of Improving RF Shielding
Even if all the screws fixing panels together enable good contact to be made there is still the possibility that RF can get in or out. This can happen particularly at VHF and above where the distance between the fixings can become a significant fraction of a wavelength. If this happens it can seriously degrade the screening. As a general rule, the distance between the fastenings should be no more than a tenth of a wave-length.

In addition, if there are any hinged panels they should be securely earthed. This can be done by connecting an earth braid between the panel and the main chassis.

Earth Braids
There are several occasions when an earthing braid is required, e.g. to earth a hinged panel or to connect a piece of equipment to a common earth point.

Although this braid can be bought, it is often easier to make it up from the screening braid from a suitable length of discarded coax. The outer PVC covering can be removed by making a small cut along its length and then stripping it back to reveal the braid. Then the braid can be removed by pushing one end of it so that its diameter increases and it can be moved along the dielectric spacing between the inner and the outer. Once removed, the braid can be pulled taut and flattened out ready for use.

4 COMPONENT IDEAS

Most radio amateurs have a junk box in some form or another. It can be a gold mine of useful bits and pieces ready to be used in the latest project and saving the time and expense of having to go out and buy new components. Sometimes the components out of the junk box may not be exactly right or there may not be a full data sheet available for them. At times like these, it can be useful to have a good knowledge of components, which ones are likely to work and which ones probably will not.

A knowledge of components can also be very helpful in knowing how best to do something. This is particularly true in respect of coils as they frequently have to be made and the circuit operation often depends heavily on their parameters. There are, of course, many other instances where a few hints and tips can be very useful saving a lot of time, effort and expense.

Resistor Colour Coding

The colour coding scheme is almost universally used for marking resistors. There are obviously some exceptions; for example some high precision resistors, some high wattage ones and the chip or surface mount resistors, if they are marked at all.

The basic system is very simple. The first two bands give the significant figures, the third indicates the multiplier and the fourth is the tolerance (as shown in Figure 4.2). So the resistor with brown (1), black (0), orange (3) and red (2) would have a resistance of 10×10^3 ohms, i.e., 10 Kohms and a tolerance of two percent.

Occasionally one sees resistors with two extra bands. They are high tolerance or precision resistors. In these cases the first three give the significant figures, the fourth is the multiplier and the fifth is the tolerance. Then the sixth band indicates the temperature coefficient which is in parts per million per degree Centigrade or Celsius. Thus a resistor with bands of red, black, green, red, green and red is a 2.05 Kohm resistor with a tolerance of $\pm0.5\%$ and a temperature coefficient of 50 ppm/°C. The main problem in decoding the value of these resistors can be in knowing which end is which and where to start

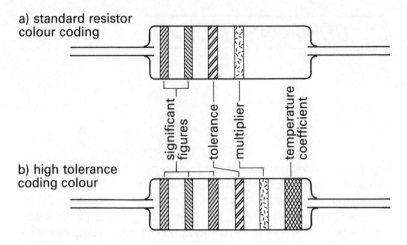

a) standard resistor colour coding

significant figures — tolerance — multiplier — temperature coefficient

b) high tolerance coding colour

Figure 4.1 Resistor colour coding schemes

reading the colours. Usually the band furthest to the right is a little wider, and failing that a little common sense helps.

Capacitor Markings

A colour coding system is also used on some capacitors. Probably the most common type to use is the polyester capacitor. For these capacitors the same basic colour coding system is used as for resistors, ie., black is zero, brown is one etc. These also have two bands for the significant figures, one band for the multiplier and one for the tolerance as shown in Figure 4.3. The main difference is that the last band is used to indicate the working voltage. Brown is used for 100 volts, red for 250 volts and yellow is used to denote a 400 volt working capacitor.

Once the capacitor value has been worked out, it may seem large. This is because it is in picofarads. In fact almost any capacitor which does not show units will be in picofarads, if it is a small value type eg., ceramic, polyester, silver mica etc. and in microfarads for a larger value type eg., electrolytic or tantalum.

Other Markings As well as colour coding, capacitors can be coded in other ways. One popular way is to use figures in a similar way to the colours in a colour coding system. This gives a code which is easily deciphered but avoids the use of a decimal point which can be rubbed off easily.

Figure 4.2 Resistor colour codes

Colour	Significant figures	Multiplier	Tolerance	Temp coefficient
Black	0	10^0		200 ppm/°C
Brown	1	10^1	±1%	100 ppm/°C
Red	2	10^2	±2%	50 ppm/°C
Orange	3	10^3		
Yellow	4	10^4		
Green	5	10^5	±0.5%	
Blue	6	10^6	±0.25%	
Violet	7	10^7	±0.1%	
Grey	8	10^8		
White	9	10^9		
Gold		10^{-1}	±5%	
Silver		10^{-2}	±10%	
None			±20%	

Figure 4.3 Colour coded capacitor

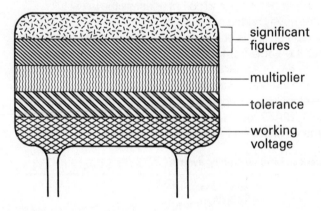

significant figures

multiplier

tolerance

working voltage

The code normally consists of three numbers followed by a letter. As might be expected the first two figures are the significant figures and the third is the multiplier. This gives the capacitance value which will be in picofarads for small value types of capacitor and in microfarads for electrolytics.

The letter which is usually present at the end of the code denotes the tolerance. This is usually either a J, K, M, or a Z. J denotes ± 5%, K is ± 10%, M is ± 20% and Z is −20% and +80%. So a ceramic capacitor marked 101J would be 10 × 10^1 or 100 pF with a tolerance of ± 5%

Figure 4.4 Numerical coding

Valve Identification

Now that semiconductors have taken over in virtually all applications, it is becoming increasingly difficult to obtain data on valves. This is in spite of the fact that there are still on the market a lot of transmitters and receivers which use valves. Often these older units operate perfectly well and offer much better value for money in spite of the fact that they may be a bit bigger and do not possess the latest frills.

It is useful to have an idea of the function of a valve without having to search for data on it. It is usually possible to gain quite a lot of information about the valve from its type number. There are two main systems which are used, giving a greater or lesser amount of data dependent upon the particular system used.

The first system, which originated in the States, applies to numbers such as 6CH6, 12AT7 etc. In this case, the first number refers to the heater voltage as shown in Figure 4.5 and the remaining letters and

Figure 4.5 American valve numbering system

1st Figure	(Indicates heater voltage)
0	Cold Cathode
1	0−1.6 v
5	4.6−5.7 v
6	5.6−6.6 v
7	6.3 v loctal
12	12.6 v
35	around 35 v

Second and other characters are type serial numbers.

Suffixed Letters	
G	Large glass envelope
GT	Small glass envelope
M	Metallised
X	Low loss base
W	Military type base

numbers form the type serial number. Thus a 6CH6 has 6.3 volt heaters, and a 12AT7 operates from a 12.6 volt heater supply. In fact in the case of the 12AT7 and several other valves (12AX7, 12AU7 etc) the heater is centre tapped so that it can operate from either 6.3 volts or 12.6 volts by placing the two halves of the heater in parallel or series. If there are any suffix letters these also tell something about the valve, for example a 6L6G has a large glass envelope and a 6V6GT has a small glass envelope.

The second system, employed by European manufacturers, applies to valves like ECC83, EABC80 etc. Using this system it is possible to tell what elements are contained in the valve. Referring to Figure 4.6 it can be seen that the first letter gives information about the heaters, and any further letters describe the various elements within the valve. Then the number indicates the base type and the particular valve type serial number.

Taking an ECC83 as an example it can be seen that it has 6.3 volt heaters (E), it contains a double triode (CC), its value base is a B9A (8) and its type serial number is 3. Similarly a PL80 has 300 mA heaters (P), it contains an output pentode (L) and has a B9A base (8).

Figure 4.6 European valve numbering system

1st Letter

A	4 v heater	C	200 mA heater
D	0.5 to 1.5 v heater	E	6.3 v heater
G	5 v heater	H	150 mA heater
K	2.0 v heater	P	300 mA heater
U	100 mA heater		

Subsequent Letters

A	Single Diode	B	Double Diode
C	Triode	D	Output Triode
E	Tetrode	F	Dentode
H	Hexode or Heptode	K	Heptode or Octode
L	Output Pentode	M	Electron Beam Indicator
N	Thyratron	Q	Nonode
T	Misc.	X	Full Wave Gas Filled Rectifier
Y	Half Wave Rectifier	Z	Full Wave Rectifier

Numbers

20–29	B8G (Loctal)
30–39	Octal
40–49	B8A
50–59	Misc.
60–69	Subminiature
80–89	B9A
90–99	B7A

If the number is over 100, subtract 100 from the number to obtain the base type.

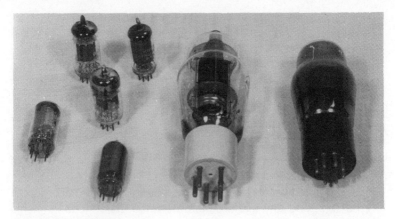

Figure 4.7 A selection of valves

Transistor Identification

In the same way that valves have a numbering system, so do transistors and diodes. In fact, there are three main systems. One is American, another Japanese and the third is European. Of the three it is the European system which is designed to give the most information about the device in its number. As shown in Figure 4.8 it consists of two letters followed by a type serial number. The first letter indicates the type of semiconductor material used in the device. The second gives the intended function and the remaining three characters give

Figure 4.8 The European (pro electron) system for transistor and diode identification

1st Letter

A	Germanium
B	Silicon
C	Gallium Arsenide

2nd Letter

A	Low Power/Signal Diode
B	Variable Capacitance Diode
C	Low Power Audio Frequency Transistor
D	Power Audio Frequency Transistor
F	Low Power High Frequency Transistor
L	Power High Frequency Transistor
S	Low Power Switching Transistor
U	Power Switching Transistor
Z	Voltage Reference Diode

Serial Number
Three figures (100–999) for devices intended for consumer applications. One letter and two figures (10–99) for devices intended for industrial equipment.

the type serial number. This serial number will be totally numeric for it is intended for consumer equipment, but if it is to be used for industrial applications the first character of the type serial number will be a letter.

Take as an example the BC107. The first letter 'B' indicates that it is a silicon device; the 'C' indicates that it is a low power audio frequency transistor and the serial number '107' having no letter included indicates that it is intended for consumer applications. Similarly a BLY33 is a silicon RF power transistor for industrial equipment.

The American or JEDEC system is quite straightforward. The first number indicates one less than the number of electrodes. This is then followed by the letter 'N' and the type serial number. Thus a 1N914 is a diode, a 2N2369 is a transistor, a 2N3819 is a JFET and a 3N211 is a dual gate FET.

Finally there is the Japanese system. This is a similar system to the American one in that it has a first number followed by a letter, in this case 'S'. A second letter indicates the type of transistor; A is PNP high frequency; B is PNP low frequency; C is NPN high frequency and D is NPN low frequency. This is then followed by the type serial number. So a 2SC137 is an NPN high frequency transistor. Most of the other systems used for current transistors are manufacturers' own 'house' codes. In some cases the type number may indicate something about the device, but in most cases it is best to refer to the manufacturer's data.

Transistor Equivalents

Circuits using transistors can be surprisingly tolerant to substitution of near equivalent components. This is partly due to the need to design into any circuit a high degree of immunity to value changes in certain parameters. Semi-conductor parameters will vary widely between different devices of the same type. To illustrate the point, have a look at the hfe or forward gain characteristic of a transistor in a data book.

The immunity to change which is designed into circuits can often be advantageous as it can enable one transistor type to be substituted for another which happens to be in the junk box. It is sometimes possible to use a device for a purpose for which it was never intended. For example, a 2N2369 can operate quite happily in some transmitter RF applications. This is because its fast switching capability gives it a good high frequency response.

When substituting like this one has to be very careful. Although it may appear that the transistor should work well, this is not always the case. One example of this occurred when trying to use a BC107 transistor having a cutoff frequency of 300 MHz in a switching circuit. The input waveform had a repetition frequency of 5 MHz and a mark space ratio of 5:1. Unfortunately the output showed no trace of the input waveform at all. Having checked that the transistor and other components were not faulty the only solution was to replace the transistor with a 2N2369. This operated perfectly well.

Very often it is possible to use almost any transistor for an application that is not particularly demanding. Beyond this it is advisable to use a specified equivalent or near equivalent device – or expect interesting problems!

Dangers of Beryllium Oxide

We are all made aware of the more obvious dangers associated with operating and running amateur equipment. The dangers of high voltage supplies, for example, are quite well publicised. However, one of the dangers which is not often brought to everyone's attention is associated with components containing beryllium oxide.

This substance has excellent heat conductance properties whilst still providing a high degree of electrical insulation. It is this rare combination of properties which makes it very attractive for use in a number of areas. One particular area where it is used is in RF power transistors where a large amount of heat has to be efficiently conducted away from a comparatively small area.

When contained within a transistor encapsulation it contains no health hazard. However, if this encapsulation is broken for any reason, it can become dangerous. Therefore, if components likely to contain beryllium oxide are to be removed from a board, great care should be taken not to damage the encapsulation. The amount of heat and force applied to the component should be kept to a minimum. If by any chance any beryllium oxide does escape and any is thought to have been inhaled or entered a cut then medical attention should be sought.

Choosing Transformers

A circuit which is common to most pieces of equipment is a power supply, and this is often added on as an afterthought. Admittedly, power supply design has been simplified with the introduction of integrated circuits. Devices like the 78-- series of regulators are cheap, easy to use and operate very well. Probably the biggest headache now when building up a PSU is the transformer. Sometimes there is one in the junk box which might fit the bill, but it is necessary to ascertain properly whether or not it is likely to be suitable. By applying a few simple rules it is quite easy to calculate whether the transformer will be suitable for the job.

In order to assess what voltage the transformer is required to deliver, all the losses have to be added up and then added to the regulated output. The first, and probably most obvious loss is the voltage drop across the regulator which has to be there for it to operate correctly. If this voltage is not maintained the regulator will 'drop out' and cease to regulate thus causing hum to get through. For supplies up to about 15 volts or so this should be no less than 2 volts, and preferably more if good regulation is required.

The second loss, if it can be called that, is the ripple. The 2 volt drop for the regulator must be the minimum at any time, so any ripple must be added on to this. Normally one should allow for about three volts of ripple on a supply such as the one shown in Figure 4.9.

Incidentally there is a useful rule of thumb for smoothing capacitors which recommends using about 200 µF for every amp the supply is required to give. If this is done then there should be about 3 volts of ripple. Larger capacitors can be used to reduce the ripple – eventually a large increase in capacitance will result in only small decreases in ripple.

There are two further losses. There is the voltage drop across the rectifier. If a bridge rectifier circuit as shown in Figure 4.10 is used then the drop will be about 1.4 volts (ie. two diode drops) whereas if a centre tapped transformer is used with the configuration shown in Figure 4.11 then the drop will be only about 0.7 volts (ie. one diode drop). The other loss is the internal drop in the transformer and this will, of course, vary with the load presented to it. Most manufacturers specify a regulation factor.

When estimating all of these losses it is well worth being pessimistic and using the regulator to remove the excess voltage. This will mean, however, that an adequate heatsink has to be used.

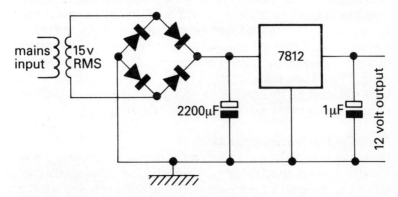

Figure 4.9 A typical regulated power supply circuit

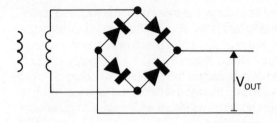

Figure 4.10 Full wave rectifier circuit using a bridge rectifier

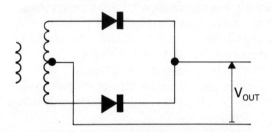

Figure 4.11 Full wave bridge rectifier circuit using a centre tapped transformer

Therefore the peak voltage produced by the transformer should be the sum of the regulated output voltage, regulator drop, peak to peak ripple, rectifier drop and transformer loss. As transformers are specified in RMS output voltage, the peak voltage delivered will be 1.4 times the RMS voltage.

Having calculated whether the voltage rating is sufficient it is still necessary to see if it will provide sufficient current. Unfortunately, this is not as simple as it may seem at first sight. A transformer rated at 1 amp will not be man enough to supply a regulator circuit delivering an amp. The reason for this is that the rectifier will only pass current when the transformer voltage exceeds the capacitor voltage (indicated by the 'smoothed output' in Figure 4.12). Then it will take a large amount of current when the capacitor is charging and the voltage across it is rising. This will take place for only a fraction of the cycle, thus giving a large current pulse for a short time. As a general rule the transformer should be derated to 0.6 or 0.7 of its specified rating.

Ripple Currents in Smoothing Capacitors
Most people know that if a capacitor is connected the wrong way round in a power supply it will fail – probably in a rather spectacular fashion, spraying tin foil everywhere. This is not the only way to make capacitors fail. Application of an excessive surge voltage, excessive

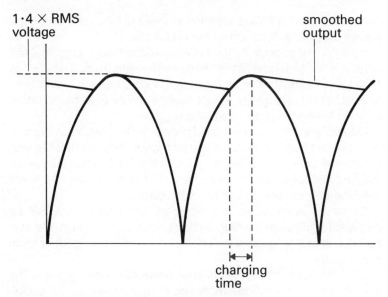

1·4 × RMS voltage   smoothed output

charging time

Figure 4.12 Waveforms in a full wave rectifier circuit

temperature and too much ripple current can all damage the capacitor.

Often the last point, that of ripple current rating, is forgotten and sometimes exceeded. Whilst in many situations such as bypass or coupling applications it may not be a problem, it most certainly has to be considered when designing power supplies.

On most of the larger electrolytic capacitors intended for use in power supplies there is a maximum ripple current rating. This will usually be stated on the can but, if not, it will most certainly be mentioned in the manufacturer's data sheet. If this value is exceeded then the capacitor will get hot and its life will be considerably reduced. In certain cases it may cause it to fail.

Unfortunately the ripple current is not just the value of the current drawn by the load. It can be greater than this because of the way in which the supply operates. The exact value of the ripple current will depend on the circuit values. So, when using an electrolytic capacitor for smoothing a supply it is well worth using one with a generous ripple current rating.

Use and Abuse of NiCads

Looking through the advertisements in the magazines, and listening on the air, one cannot help but notice how much portable equipment is being used, especially on VHF and UHF. Most of this is powered by

Nickel Cadmium or NiCad batteries as they are the most economical source of power in spite of their high initial cost.

If they are used properly these batteries can have a long life, but it is quite easy to shorten their life by mistreating them. This can be costly, even if they can be replaced by standard types of NiCad, but if special battery packs have to be bought it can be even more expensive. So it is worth taking care of them.

Generally a cell is said to have finished its life when the maximum charge it can hold is only half of its rated value. Normally it takes several hundred charge/discharge cycles for this to occur, but this can be reduced dramatically by mistreatment. The usual causes of this are overcharging, reverse charging and temperature.

As far as temperature is concerned, the performance will be degraded particularly by high temperatures, but low ones are also harmful. In fact a NiCad will perform best and last longest at room temperature.

Most failures in NiCads result from incorrectly charging them. The manufacturer's instructions regarding charging them should be followed carefully. Excessive charging rates will soon degrade the performance, although there are some types of cell which are designed to be charged quickly. Lower charging rates will not lead to any damage but it will obviously take longer to charge, and this may be inconvenient.

Overcharging is another problem. In fact there are two ways in which a cell can be overcharged. The first is a slight but prolonged overcharge. This manifests itself after the charging has finished by a reduction in the output voltage. Fortunately this can be rectified to a degree by giving the cell a deep discharge.

The other form of overcharge occurs when the cell is charged at the normal rate for too long. When this happens the cell can become hot and gases are produced which escape through a vent. This obviously reduces the amount of active chemicals within the NiCad, and this leads to a reduction in its capacity. One of the best ways to overcome this problem is to discharge fully each cell before charging and then use a charger with a built-in timer. In this way the correct amount of charge will be given to the cell.

When charging cells it is perfectly permissible to charge them in series as the charging process is dependent upon the amount of current passing through each cell. In fact, if several cells are to be used together, it is advisable to do this as each cell will receive the same amount of charge.

Cells should not, however, be charged in parallel as their characteristics will be slightly different and each one will pass a different amount of current. This could mean one cell will be overcharged whilst another is left undercharged.

If a battery needs to be completely discharged, each cell should be discharged separately because each one will hold a slightly different amount of charge. If they are discharged as the whole battery, then, as it nears discharge, some cells will have a little charge left whilst others will have none. This will mean that some cells will actually become reverse charged, another condition which causes the release of gases and damage to the cell.

One interesting effect which was first detected in the NiCad battery packs used in space craft is the 'memory effect'. This occurs when a cell is repeatedly only partially discharged. The cell appears to memorise the amount of charge which is normally drawn and after this

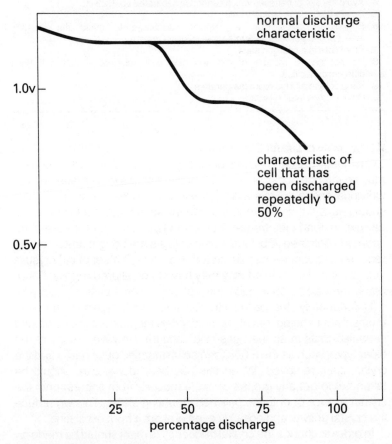

Figure 4.13 The memory effect

point the voltage drops. This effect can usually be overcome by giving the cell a few charge/discharge cycles and leaving it to become fully discharged each time.

When storing cells for any length of time without use it is best to leave them fully charged. The reason for this is that if they are left discharged internal shorts are likely to be caused. If they are left fully charged this is less likely to happen.

Finally Figure 4.14 summarises some rules for getting the best out of NiCads.

Figure 4.14 Ten rules for nicad users

1. Do not overcharge cells.
2. Do not reverse charge cells.
3. Do not completely discharge cells connected in a battery – discharge them separately.
4. Avoid the use of batteries which cannot be separated eg. PP9 etc.
5. Do not fast charge ordinary cells.
6. Periodically gives cells a complete discharge to avoid the memory effect and to get cells back to a known state of charge.
7. Store cells in a charged state.
8. Do not use a mixture of old and new cells together – they will have different capacities.
9. Keep cells at around room temperature.
10. Do not short circuit a NiCad.

A New Lease of Life for Old Crystals

Even though crystal filters are generally bought ready-made, there still may be the odd occasion when a crystal filter has to be made up from individual crystals. If this is done, it will be found that not all the crystals should have the same frequency otherwise the bandwidth will be too narrow. Instead their frequencies should be such that the correct response is obtained. Obviously if the crystals are bought new, then the required frequencies can be stated. However, if a set of old crystals from the junk box is used they may have to be altered slightly. Fortunately this can be done quite successfully in some cases.

Unfortunately, the newer HC18U and HC6U types do not lend themselves to being 'got at' as the crystals themselves are contained in sealed units in an inert gas or a vacuum. However, many of the older types such as the FT243 can be dismantled quite easily and the crystal itself removed. When this has been done, care should be taken not to get any grease or dirt (especially from fingers) onto the crystal surface or onto the polished mounting plates. This will reduce the crystal activity and possibly even prevent it from oscillating.

In order to change the crystal frequency, a mark should be made on the crystal surface using a soft lead (B or 2B) pencil. The actual amount of pencil 'lead' to be added to the crystal surface will have to be determined by experiment because the amount required will vary between one crystal and another. It is advisable to put a little on each time and measure the effect on a test oscillator. This can be repeated until the correct frequency is obtained. When doing this, beware of trying to obtain too much change in frequency as the activity will fall after a certain point.

If the crystal activity falls, or fingers stray onto it, then it is possible to clean the surface. This can be done quite successfully with almost any organic solvent which can remove grease.

Coils

One type of component which is avoided wherever possible in electronics is the coil. Although a wide selection of chokes are available 'off-the-shelf' many coils still have to be wound individually and this is both time consuming and expensive. In spite of this coils are still included in many designs as there are no substitutes for them at radio frequencies. Because of this they find a lot of use in amateur radio circuits, and the home constructor often finds himself winding coils and building them into projects.

Coil Winding Wire

When winding coils it is convenient to use the self-stripping type of enamelled wire. The idea of this is that when the soldering iron is applied to the wire, the enamel burns off leaving the copper which is then tinned by the solder on the iron, leaving the end of the wire ready for soldering. Take care to ensure that bad joints are not made.

The enamel, however, does not always burn off as easily as one may imagine. A reasonably hot iron is needed and this has to be held on the wire for a little while. It helps if some of the enamel can be scraped off before soldering as this enables the iron to make better thermal contact with the wire and burn the enamel off faster.

Once this has been done, check that the wire is tinned properly over the entire area that is required for soldering as it is possible to leave small areas of enamel still on the wire. If the wire is soldered with enamel still on it, there is the risk of making a bad joint.

When the coil is soldered in, it is worth checking for continuity between the soldered joints. This may seem a little over-cautious but it is sometimes not possible to detect visually a bad joint and it could take a long time to discover the problem later.

Finally, this type of wire should be soldered in a well-ventilated area as excessive amounts of the fumes from the enamel can be harmful.

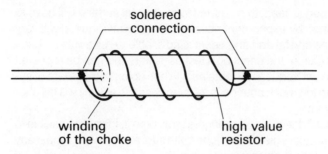

soldered
connection

winding
of the choke

high value
resistor

Figure 4.15 Using a resistor as a former for small value chokes

A Resistor Former for Small Chokes

Small chokes of a few turns can come in handy on many occasions. They can often be used as anti-parasitic chokes in valve amplifiers or in semiconductor circuits to prevent unwanted RF pickup. They can be made up quite easily by winding a few turns of wire to the resistor leads as shown in Figure 4.15. In this way the resistor acts as a former as well as giving mechanical strength to the whole assembly.

Normally when making an anti-parasitic choke for a valve amplifier, a resistor with a value of 100 K ohms or higher is best. In fact, provided that its resistance is reasonably high, its value is of no importance as it will have a negligible effect on the choke. Usually a half watt resistor will provide a sufficiently large former although for high power amplifiers a larger one may be needed. Once a suitable resistor has been chosen, three or four turns of wire can be wound round the resistor and this should be sufficient to suppress most VHF parasites.

If the choke is to be used for semiconductor circuits, a quarter watt resistor can be used. This should be made up in the same way as before. If necessary a few extra turns can be used, but this will obviously need a thin gauge of wire.

Advantages of Torroids

Torroidal chokes, coils and transformers are appearing more often in circuit designs these days. For example, one only has to look at the majority of QRP transmitter designs to see a liberal use of torroidal inductors, particularly in the output filters.

Torroidal coils are a little more difficult to wind than the more traditional style ones on a straight former and the cores are a bit more expensive. So what is the advantage? Basically, they require little or no shielding. As they are circular, or torroidal, all the magnetic flux lines are contained within the core. This means that the coil will not have any effect on the one next to it. The great advantage of this is that

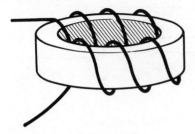

Figure 4.16 A torroidal inductor

circuits using torroids do not require metal shielding cans for the coils. Circuits using torroids also tend to be more stable because there is almost no cross talk between one coil and the next.

Ferrite Bead Chokes

One very convenient way of making fairly small value chokes is to use ferrite beads. Apart from the fact that it is easy to make chokes in this way, they are ideal because they are torroidal. This means that there is less likelihood of interaction with other inductors with the resultant possibility of instability.

One popular formula for chokes for VHF circuits is two and a half turns on an FX1115 bead. This works very well in many circuits, but there is little idea of what the inductance actually is. In fact one turn on one of these beads gives around 1 μH. This value is approximate because the permeability of the material (which determines the inductance) will vary from core to core. In addition, the inductance will vary with frequency. However, it gives a good idea of what the inductance will be.

Having obtained this value for the inductance of a single turn, it is quite simple to calculate the inductance of a choke with several turns by using the formula:

$$\frac{N_1}{N_2} = \sqrt{\frac{L_1}{L_2}} \qquad \begin{array}{l} N = \text{the number of turns} \\ L = \text{the inductance} \end{array}$$

From this it can be seen that two and a half turns on an FX1115 bead gives an inductance of around 6 μH – quite sufficient for most applications at 2 metres. Other values can be wound for other purposes. However, it should be remembered that for slightly higher current applications the maximum current has to be reduced as the number of turns is increased. This has to be done to ensure the core does not saturate.

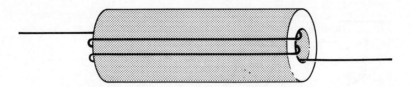

Figure 4.17 A typical ferrite bead choke

Ferrite Beads to Suppress Parasitics

Ferrite beads need not always be used as small formers for torroidal chokes. They can also be used to add just a small amount of inductance to a lead by passing a wire through the centre of the bead. This idea can often be used to stop parasitics in the same way as a grid, gate or base stopper resistor. However, it has the advantage that it is a very easy addition to make, as shown in Figure 4.18. If the parasitics can be stopped in this way, it saves having to add extra components which may require an existing printed circuit board to be changed.

Inductance of a Single Layer Coil

It is an unfortunate fact of life that a lot of very useful circuits appearing in the magazines only give an inductance value for a coil and omit the wiring details. Alternatively, the design may be from abroad and the coil former may not be available. Either way, it can be a very time consuming occupation to determine a suitable set of winding details by trial and error.

Fortunately, it is possible to calculate the inductance from several formulae which are available. Some of these formulae can be quite complicated as the inductance of a coil depends on a large number of parameters. Normally an exact value is not required because it is pos-

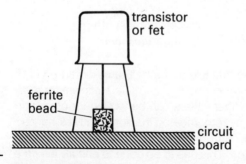

Figure 4.18 Using a ferrite bead to suppress spurious oscillations

sible to obtain a large degree of tuning using a variable core. In view of this a simple, usable formulae to give a good starting point is all that is needed.

In order to simplify, the formulae have been divided into three for different length to diameter ratios.

$$N = 10\sqrt{\frac{L}{D}}$$ for winding where length is twice the diameter

$$N = 7.6\sqrt{\frac{L}{D}}$$ for winding where length equals the diameter

$$N = 6.2\sqrt{\frac{L}{D}}$$ for winding where length is half the diameter

In these formulae N = number of turns, L = inductance in microhenries and D is the diameter in inches.

As these formulae only give a good starting point, a little experimentation may be needed afterwards. It should also be remembered that, if a ferrite core is used, it will give a significant increase in inductance, possibly up to twice the original value. This will obviously be dependent on factors like the permeability of the core, its length and so forth. Alternatively a brass core can be used if one is available to reduce the inductance.

5 CIRCUIT IDEAS

An enjoyable aspect of amateur radio is building up circuits and experimenting with them. It is quite true that it is possible to gain more pleasure using a homebrew circuit rather than just operating a piece of ready built equipment, even though it may have all the latest gadgets and operating aids. There are also the economics to be considered. It is sometimes possible to build up a circuit for less than it would cost in the shops. With the price of today's equipment this can become very important.

Although there are a large number of published circuit designs for complete projects, this chapter aims to provide a source for useful ideas. Some of them may be incorporated into other circuits to improve them or add extra facilities. Alternatively others may save time and money or even just help to get over a problem.

A Home-Built Dummy Load

With the wide availability of many off-the-shelf ancillary pieces of equipment, it is very often easy to forget that some of them are quite easy to construct. One such item is a dummy load. This can be built from a number of small one watt or half watt resistors wired in parallel.

In order that the load should operate satisfactorily it is necessary to avoid introducing any more inductance than is absolutely necessary. Wire wound resistors must not be used as they become almost totally inductive at radio frequencies. Also the arrangement of the resistors needs a little thought. The resistors can be arranged either in a circular formation as shown in Figure 5.1 or in the form of a matrix as shown in Figure 5.2. Whilst the circular formation is ideal for a small number of resistors, the matrix is probably better as the numbers increase. In fact, the one in the diagram used 100 5K1 one watt resistors to give a hundred watt 51 ohm load. Although not exactly 50 ohms, the resistance was close enough to the required value not to cause any problems.

When constructing the load the leads should be kept as short as possible (within reason) in order to reduce the inductance. However, there should also be sufficient space left between the resistors for

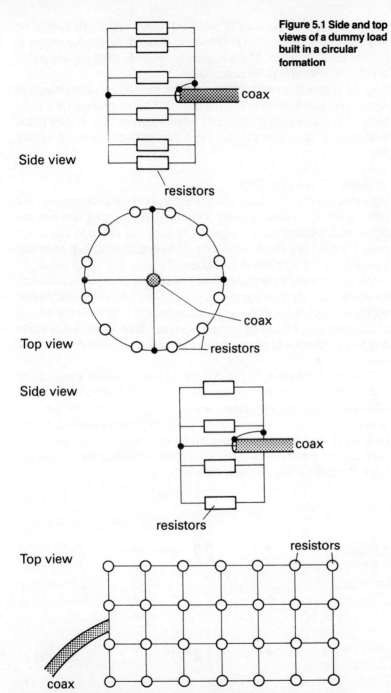

Figure 5.1 Side and top views of a dummy load built in a circular formation

Side view

coax

resistors

Top view

coax

resistors

Side view

coax

resistors

Top view

resistors

coax

Figure 5.2 Side and top views of dummy load built in a matrix formation

ventilation and, if the load is mounted in a case, there should be adequate air flow to avoid overheating. If possible, it is advisable to put the load into a box. This will prevent anybody touching any points which might have high RF potentials.

Once in operation the prototype load performed well. It handled 100 watts quite comfortably and gave a VSWR of better than 1.2:1 at 30 MHz. The performance obviously deteriorated at VHF in view of the lead lengths, but it proved to be more than satisfactory for HF operation.

A Simple TV High Pass Filter

Now that the old VHF television transmissions have ceased and the UHF spectrum is used instead, TVI from HF transmissions has become less widespread. However, this does not mean to say that it does not exist any more, especially if one lives in a built up area with transmitting aerials close to the house.

Very often the interference arises not as a result of harmonics from the transmitter but from the television front end becoming overloaded by the very strong local amateur transmissions. When this is the case it is usually easy to cure by using a high pass filter. Whilst it is usual to buy a filter from the local emporium it is quite an easy matter to make one.

The circuit shown in Figure 5.3 has proved useful in a number of instances and it is quite easy to make. The coils consist of 2 turns of 20 swg copper wire (air spaced) with a 1/4 inch internal diameter and wound over a length of about 1/4 inch. The capacitors should be good UHF types and lead lengths should be kept as short as possible to minimise the insertion loss. This becomes particularly important if the higher frequency TV channels are used.

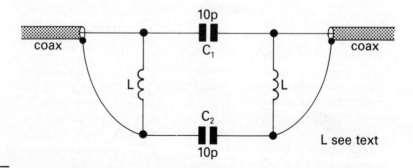

Figure 5.3 Circuit of high pass filter

Once constructed, the filter can be housed in a small case. Although it is possible to obtain suitable project boxes for these filters they are often expensive. An alternative is to use an old tobacco tin. These are fine, provided that one does not mind 'Old Holborn' or the like being advertised at the back of the television.

Overcoming Hi-fi Breakthrough

Unfortunately TVI is not the only form of interference which can be caused to domestic equipment. Today's modern hi-fi systems are also prone to RF pickup from nearby transmitters. The long leads required for spacing the speakers far enough apart for good stereo act as good aerials, especially if they happen to be resonant on an amateur band. This is, of course, not out of the question if the band in use is ten or fifteen metres.

When the speaker leads to pick up RF it enters the amplifier at its output, and it is often carried back to the earlier stages of the amplifier along the negative feedback line. After being rectified by any slight non-linearity in one of these stages it is amplified, appearing at the output as a disturbing noise for the unsuspecting neighbour.

One way in which this can be cured is to place a small inductor in the speaker lead. The inductor must have a very low resistance so that it does not impair the performance of the speaker. The best way to achieve this is to wind a few turns of the speaker lead around some ferrite. One suitable former could be made from a section of ferrite salvaged from the aerial of a discarded portable radio. Better still, it could be wound onto a torroidal core.

A Simple Mains Filter

It often happens that interference to domestic equipment can be caused by RF from the transmitter getting onto the mains. This is particularly likely to happen if an unbalanced aerial, such as a long wire, is used. As the aerial needs an earth or counterpoise for it to operate, the mains wiring can become part of this system and can carry quite a lot of RF. Whilst it may seem that removing the earth wire to the transmitter may solve the problem, this is very unwise from the point of view of safety. It is also quite possible that the live and neutral wires will be carrying RF as well and so little may be gained.

Fortunately there are ways round the problem. One is to construct a proper low pass filter but this can be both time consuming and expensive. It is much cheaper and quicker to make a simple torroidal choke similar in many respects to that used for the speakers in the hi-fi system. The ferrite ring itself will obviously have to be larger to accommodate the bigger wire, but in other respects it will be the same.

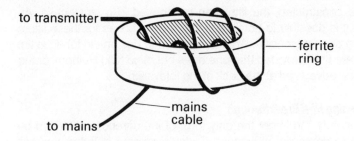

to transmitter

ferrite ring

mains cable

to mains

Figure 5.4 Ferrite ring mains filter

The wire should be wound around the ring a few times, as shown in Figure 5.4. Also the torroid should be placed as close to the transmitter as possible to stop the RF as soon as it comes out of the transmitter.

A Sure-Fire Crystal Oscillator

Crystal oscillators come in useful for a variety of jobs around the shack. They can be used for crystal calibrators, frequency references in synthesisers, oscillators in receivers and also for transmitters, in particular QRP transmitters.

The circuit shown in Figure 5.5 has been used frequently and has always given good results. With the values shown it has operated satisfactorily from below 3.5 MHz to 15 MHz or more, and then by increasing C_2 and C_3 its range can be extended down in frequency. The variable capacitor is included to trim the oscillator to the exact frequency which is required. In fact, if a coil is placed in series with the crystal the frequency can be pulled further and used in a QRP transmitter. Some experimentation is required to find the optimum results but, when doing this, it should be remembered that if the coil is made too large or the capacitance too small then the circuit may become unstable or its output may fall.

The circuit is quite tolerant of supply voltage and operates quite satisfactorily between about 9 and 15 volts. It also gives a fairly substantial output of about 2 volts peak to peak when it is not loaded.

Easy Way of Obtaining FM

There are two main ways of generating frequency modulation using a crystal oscillator. One is to place a varactor (or varicap) diode either across or in series with the crystal. The other is to add a phase modulator after the oscillator. Both of these methods have their drawbacks. The use of a varactor diode in the crystal circuit is not always

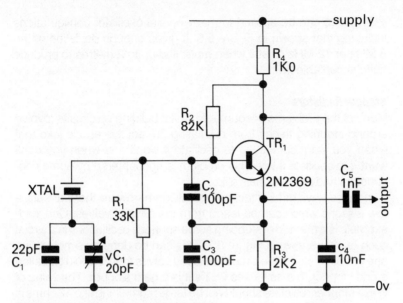

Figure 5.5 Sure fire crystal oscillator circuit

ideal as the characteristic of the diode is non-linear and will introduce distortion. Also, the amount of deviation obtained will vary between one crystal and the next making it difficult to obtain consistent results if a bank of crystals are used. The better method is to use a phase modulator after the oscillator. Although this gives superior results it requires more components and adds to the complexity of the overall circuit.

Another way of obtaining frequency modulation, which is both easy to use and produces acceptably good results, is simply to apply the audio to the base of the oscillator transistor. This should be done using a filter as shown in Figure 5.6. This not only prevents the RF from getting back down the audio line, but it also stops the audio line from loading the oscillator and altering its operation.

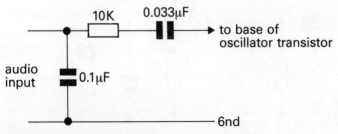

Figure 5.6 Filter used for frequency modulating a crystal oscillator

The idea can be applied to many crystal oscillator configurations including that shown in Figure 5.5. It gives enough deviation on an 8 MHz or 12 MHz crystal when multiplied up to 2 metres to produce quite acceptable FM.

Stopper Resistors

Many of the problems encountered whilst building up circuits revolve around stopping things from oscillating. In fact it is an old joke that when you want something to oscillate it won't and when you don't want it to oscillate it will! This problem is encountered by home constructors and professionals alike.

In these days of high technology silicon hardware there are still a few lessons which can be learnt from the days of valves. One such example is a method of stopping some spurious oscillations in a circuit such as that shown in Figure 5.7. This can be done by simply using one extra resistor. If the circuit had used valves then it would be called a grid stopper, but as it uses FET's it is a gate stopper. The value of these stopper resistors should not be large; something like 47 ohms is a good start. This should enable the oscillation to be stopped without impairing the operation. Obviously if the resistor has to be increased the gain of the circuit will fall, especially at high frequencies where the gate channel capacitance has to be taken into consideration. Also, if very large values have to be tried then it is a safe bet that the oscillation involves another part of the circuit and another method should be tried.

Although this example shows a gate stopper resistor in a FET circuit, it can just as easily be used with ordinary bipolar transistors. In this case a small resistor in the base can prove to be quite effective.

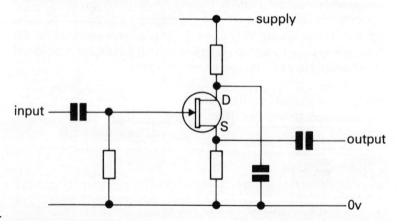

Figure 5.7 Unmodified FET source follower

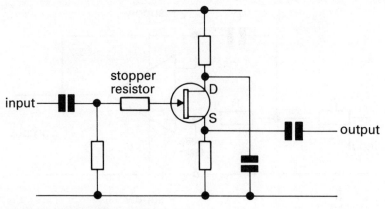

Figure 5.8 Modified FET follower circuit

Active Filters Using Op-Amps

There are a large number of applications where low pass or high pass filters are needed. Some of them are for filtering out unwanted signals at radio frequencies. Others are for use at audio frequencies and include applications in direct conversion receivers, or any receiver needing some audio filtering; speech processors where both low and high frequencies need to be used; and many other instances.

For these audio applications, the possibility of using coils in the circuit becomes rather impracticable as the values and the resultant coils are large. Fortunately it is not necessary to use coils, as it is possible to make an active filter to give the same performance using an amplifier with some resistors and capacitors in the feedback network.

The circuits for an active low pass filter and an active high pass filter are shown in Figure 5.9 and Figure 5.10. Operational amplifiers have been used in each case as they have a very high input impedance and low output impedance. This enables them to load the filter network less and give results almost identical to the calculations without having to worry about loading effects.

The circuits which are shown are known as two pole filters. They give an ultimate roll off of 12 dB per octave, or 40 dB per decade, whichever way is the more convenient to express it. If more rejection is required the two or more stages can be cascaded.

Whilst the calculation of the values can be complicated it is possible to simplify them if some restrictions are placed on the design. To do this a Butterworth type filter response has been chosen. This gives a flat response within the passband, but it does not reach its ultimate roll off as quickly as some other types. In fact for most audio applications this is the type of response which is wanted.

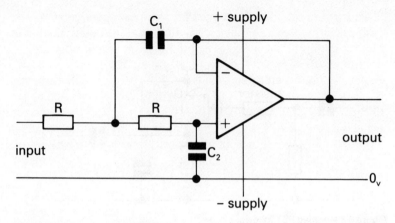

Figure 5.9 Low pass active filter

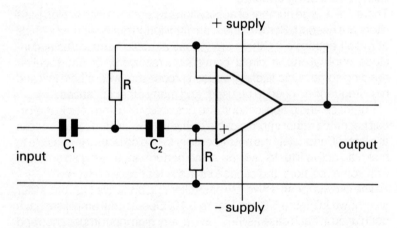

Figure 5.10 High pass active filter

The formulae needed to calculate the filter values is quite simple and applies to both filter circuits:-

$$f = \frac{\sqrt{2}}{4\pi RC_2}$$

$$C_1 = 2C_2$$

f = cut-off frequency ie., frequency when the response has fallen to −3 dB.

Usually the cut-off frequency will be the known parameter leaving both resistor values as unknowns. Therefore it is quite normal to choose a convenient value for one component and to calculate the others from the formulae. For example, if a cut-off frequency of 3 KHz is required the R could be chosen to be 24 KΩ. This would mean that C_1 would be 1563 pF or to take the nearest preferred value 1500 pF. C_2 would then be 3000 pF. These values could then be applied to either circuit to give a low or a high pass filter with a 3 KHz cut-off frequency. If the value of a capacitor or resistor turns out to be inconvenient then it is a simple matter to choose a new starting point and repeat the calculations until suitable values are found.

a) low pass filter

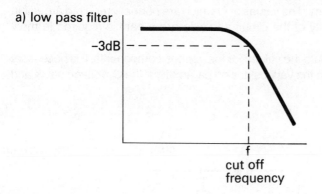

b) high pass filter

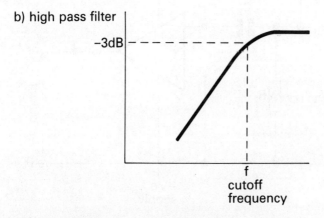

Figure 5.11 Frequency response of high and low pass filters

Active Bandpass Filters

It is possible to make not only low pass and high pass filters using operational amplifiers, but also bandpass ones. These filters again have to be restricted to audio frequencies or a little higher, but they do come in handy for a variety of jobs. For example, they can be used for CW audio filters as they are very easy to add in to existing receivers or include in new designs. Another use is as a tone bursts filter as part of a tone burst detector.

The basic circuit is shown in Figure 5.12. Again it is fairly straightforward because this configuration has only one operational amplifier. The only other components which are required are a handful of capacitors and resistors.

Obviously the simplicity of the circuit does mean that there are a few limitations. The Q and gain both have to be kept to fairly low values. However, for the majority of uses they should not cause any problems. The values of Q which are normally required are within the capability of the circuit and little or no gain is required in most instances.

To calculate the values for the various components, it is necessary to decide on the various design parameters: the Q, voltage gain[G] and

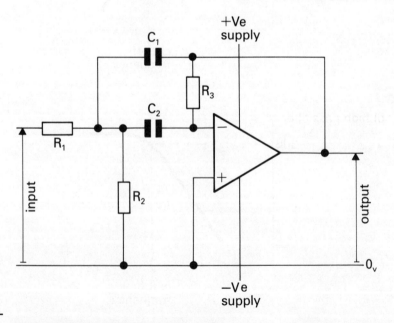

Figure 5.12 Basic circuit of the bandpass filter

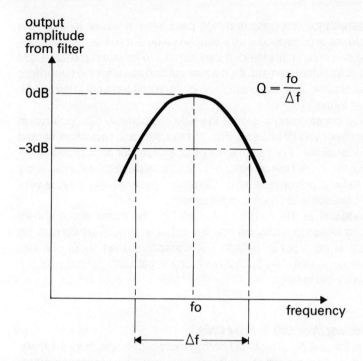

$$Q = \frac{f_0}{\Delta f}$$

Figure 5.13 The frequency response of the bandpass filter

centre frequency f_0. For many CW filters a centre frequency of 750 Hz is ideal, and as no voltage gain is required, G can be set to 1. The next parameter which has to be set is the Q and this determines the bandwidth as shown in Figure 5.13. In this case a Q of 5 would give a bandwidth of 150 Hz.

Once these figures have been set the filter components can be calculated from the formulae below:

$$C = C_1 = C_2$$

$$R_1 = \frac{Q}{2\pi f_0 C G}$$

$$R_2 = \frac{Q}{(2Q^2 - G)\, 2\pi f_0 C}$$

$$R_3 = \frac{Q}{\pi f_0 C}$$

Just as with the low pass and high pass filters it is best to choose a suitable value of capacitor as a starting value. From this, all the resistor values can be calculated. If they turn out to be much too large, too small, or just inconvenient, then a new starting value for the capacitors can be chosen and the resistor values recalculated until suitable values are found.

When constructing a filter of this type, reasonably close tolerance components should be used. Both the resistors and capacitors should be 5% or better. For the resistors this should not be a problem, as most of them are now at worst 5%, but the capacitors will most likely have to be a polystyrene type. Ceramic types normally have a very wide tolerance and should not be used.

In addition to the components used on the basic circuit shown coupling capacitors may be needed to block any DC which may be present at the input or output. The capacitors used should be sufficiently large that they do not have any significant reactance at the frequency being used.

Suppressing Back EMF in Relay Coils

Even in these days of high technology there is still a place for the lowly relay. It is often difficult to achieve a very low 'ON' resistance and a very high 'OFF' resistance with semi-conductors. In cases where both of these are needed the relay comes in very useful. However there is one point to be taken into account, i.e. the back EMF generated by the relay coil when the current is switched off. If this is not suppressed it can have a disastrous effect on any semiconductor devices around it.

A couple of typical circuits using relays are shown in Figure 5.14. The diode across the relay performs the transient voltage suppression. When the relay is ON and the current is flowing through the coil, the diode will be reverse biased and it does not affect the operation of the circuit. However when the current stops flowing through the coil a back EMF is set up. This causes the diode to become forward biased, and a current to flow through it. In this way the back EMF is dissipated quite harmlessly and the voltage spikes are avoided.

If the diode were not present a large EMF could have built up. This could either cause arcing across a switch or, in the case of the transistor circuit, it could destroy the device.

In practice, the choice of diode is not particularly critical. It will need to have a peak inverse voltage rating in excess of the relay operating voltage. Also it will need to be capable of handling enough current to dissipate the voltage spike. The highest this will be is the same as the operating current of the relay. So in most cases a small signal diode

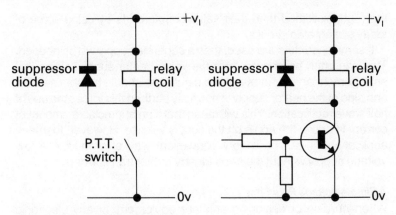

Figure 5.14 Two circuits using relays

such as a 1N914 or 1N4148 will suffice but, if a larger relay is used, a diode in the 1N4000 series is probably a safer bet.

Powering Semiconductor Circuits from Valve Heater Supply

Transmitters and receivers using valves represent very good value on the second hand market. Unfortunately one of the drawbacks is that it is not always easy to add modifications involving more valves. The extra space, power, and on top of this the difficulty of the metalwork often prevent the modifications from being worth while. However, it can be quite convenient to hybridise the equipment by using semiconductor circuits. This approach has several advantages: the circuits are smaller, they consume less power, and they don't require too much metalwork as a rule. In addition, the circuits can be made

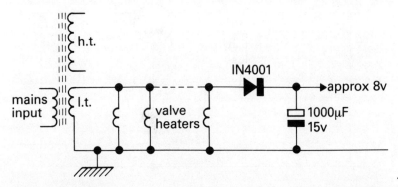

Figure 5.15 Simple supply for semiconductor circuits from a valve heater supply

more sophisticated than their valve counterparts by using some of today's integrated circuits.

If semiconductors are used, then a separate supply will be needed. This can often be derived from the valve heater supply. Using this arrangement an output of around eight volts should be obtained. As one side of the heater supply is normally earthed this forces the use of half wave rectification. This will mean that if an appreciable amount of current is drawn the ripple on the supply will rise. However, for many applications it can be a very convenient way of producing a low voltage supply without the need for any further transformers.

A Simple Voltage Regulator

When building or designing amateur equipment, or any electronic equipment for that matter, there is often a need for a rough and ready form of voltage stabilisation. Whilst it is very easy to use one of the fixed voltage regulators such as the popular 78-- series, it is an unfortunate fact of life that the right one is never in the junk box. Fortunately, it is quite easy to construct a simple regulator circuit using only a resistor, transistor and a zener diode as shown in Figure 5.16.

The way in which the circuit operates is quite simple. The resistor and zener diode provide the voltage reference and the transistor is connected as an emitter follower to increase the current capability. The resistor R_1 is chosen so that the zener diode draws 10–15 mA. This will keep it conducting sufficiently for its voltage to remain fairly constant whilst keeping the heat dissipation in the zener within limits for most cases. Using Ohms Law the resistor value can be calculated. The voltage across the resistor can be found by simply subtracting the zener voltage from the input voltage. The resistor value is then the

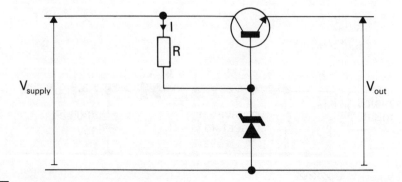

Figure 5.16 A simple regulator circuit

voltage across the resistor divided by the current into the zener diode. In fact, the base of the transistor will take some current depending upon its load, but the resistor-zener network should be able to accommodate this without losing stabilisation, provided that not too much current is taken from the output. In fact, the base current will be the output current divided by the current gain of the transistor. Therefore the circuit ought to be able to supply around 50 mA without too much difficulty.

There are obviously a few drawbacks to a circuit as simple as this. Firstly, the regulation is not particularly good and, secondly, the output voltage is 0.6 v less than the zener diode voltage. Despite these disadvantages, it is still possible to use this type of circuit in many applications.

Regulator with Adjustable Output

It is a fairly easy matter to improve the regulation of the circuit shown in Figure 5.16. If a few extra components are included, then not only does the regulation improve but the output voltage can be adjustable as well.

The circuit for this regulator is shown in Figure 5.17. It is still fairly simple and its operation is quite straightforward. TR1 forms the series pass transistor which is controlled to give the correct output voltage. It is driven by TR2 which acts as a differential amplifier which senses the voltage at point A which is a portion of the output and then compares it with the zener diode voltage. If the voltage at A is greater than that of the zener diode plus the turn on voltage of the transistor, then it will cause more current to flow through TR2, pulling down the volt-

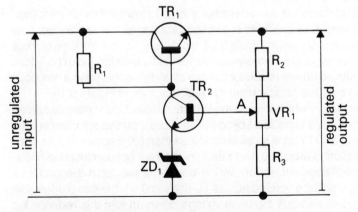

Figure 5.17 Voltage regulator with adjustable output

age at the base of TR1. This will in turn reduce the output voltage. Conversely if the voltage at A is too low the output will be raised and this forms a feedback loop within the circuit which tries to maintain a constant output voltage. It also allows adjustment of the final output by varying the fraction of the output which is compared with the zener diode. In this way the circuit provides both good regulation and an adjustable output.

The values in the circuit are not particularly critical. The value of the zener diode should be about half the required output voltage so that A should sit at about this value as well. A suitable set of values for a 12 volt supply could be 270 ohms for R_2 and R_3 and 500 ohms for VR1. R_1 is included to supply the current for the zener diode. It can easily be chosen by knowing the voltage across it and the current flowing through it into the diode. Suitable values for the 12 v supply could be 1 K ohms for R_1 and a 5 v 6 zener diode. The choice of transistors depends mainly on what is in the junk box. TR1 should be capable of dissipating some power – and could possibly be a 2N3053. TR2 need only be a small signal device like a BC107.

A Simple Current Limiter
Power supplies are always prone to having their output short circuited. A dropped screwdriver invariably comes to rest across the power lines, or a probe may slip causing a short to earth. Fortunately most of the regulator IC's, as well as commercially made supplies, have some form of current limiting built into them. However, the circuits which find their way into homebrew equipment often do not have any limiting and tend to blow if the output is shorted. Sometimes it is unlikely that they would require a current limiter, but at other times it could be very useful.

In such cases, it is not necessary to go to town with a very sophisticated limiter, as a simple protection circuit is all that is needed! One such idea is shown in Figure 5.18. It only requires the addition of three extra components, namely two diodes and a resistor. It can be added into many different regulator circuits although in this case it has been added onto the basic emitter follower regulator of Figure 5.16.

The way in which it operates is quite simple. Under normal operating conditions there is a voltage drop of 0.6 v across the base emitter junction of Q1 plus a small drop of less than 0.6 v across R_2. As each of the diodes require 0.6 v to turn on, they will not conduct under normal conditions. However, as the current drawn from the circuit increases and the voltage across R_2 rises to 0.6 v the two diodes start to conduct and pull the base voltage down. In turn this reduces the output voltage and hence the current. This means that the value of R_2

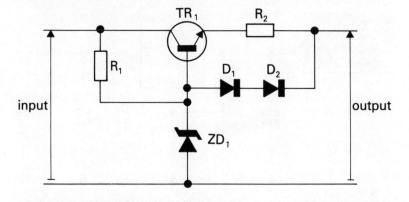

Figure 5.18 Regulated circuit with current limiting

should be chosen so that it develops about 0.6 v across it when the maximum permissible current is drawn.

The circuit does have one drawback. As R_2 is in series with the output it will cause the output voltage to fall slightly as the current is increased even when it is not limiting the current. Fortunately this is not a problem in many radio applications

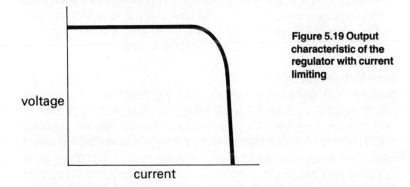

Figure 5.19 Output characteristic of the regulator with current limiting

Oscillating Regulator IC's

Despite the simplicity of the circuitry it is quite possible that power supplies built up using the 'ready made' regulator IC's will not always work. One cause can be that the IC has decided to oscillate. This problem is almost invariably caused by the fact that the smoothing capacitor and regulator are connected by a long lead. The first reaction to solving the problem is to move the two components closer together and shorten the lead. Often this is not convenient and, in

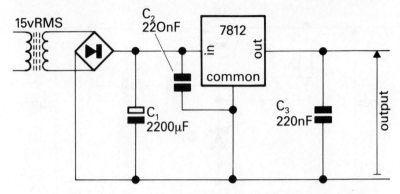

Figure 5.20 12 volt regulator using 7812 with capacitor to suppress oscillation

cases like these, it is possible to place a smaller capacitor C_2 (between 100 nF and 1μF) across the input and earth terminals as shown in Figure 5.20. This capacitor should obviously be placed as close to the IC as is conveniently possible.

Although the diagram shows a typical example using a 7812 the same idea can be used on many other regulators, and not just the 78-- series.

In addition to this capacitor, another one is sometimes seen connected across the output. This is not necessary to keep the regulator stable. Instead it is used to improve the transient response if logic circuitry is used. It may also help if there is RF around.

Over-Voltage Protection
It has already been mentioned that many regulator IC's incorporate safety features such as current limiting, over-temperature protection and so forth. However, there is always a chance that the regulator could fail in such a way that the output voltage would rise. The usual reason for this is that the series regulator has failed and become short circuited. The result is that the output voltage rises to the full voltage on the smoothing capacitor before the regulator. As this is several volts above the required output, any equipment connected to the power supply could be very seriously damaged. Even though this type of failure occurs fairly infrequently in a well designed supply its consequences are drastic. As a result many power supplies are fitted with 'over-voltage' protection.

Several schemes have been developed to overcome this problem, but the one shown in Figure 5.21 has become more widely used than the others. It operates quite simply and requires only the addition of a few extra components. Under normal operating conditions the zener

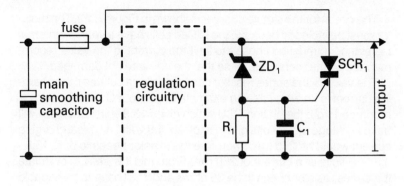

Figure 5.21 A popular circuit for overvoltage protection

diode voltage is not exceeded and no current flows through it. Accordingly the SCR gate is not triggered and the circuit has no effect on the operation of the supply. If the circuit develops a fault and the voltage rises above the zener voltage then current will begin to flow through it and the voltage will rise on the gate of the SCR. This will cause the SCR to conduct, shorting the output of the fuse to ground, which blows it and removes the supply to the input of the fault regulator. The value of R_1 is not critical, 10 K ohms being a suitable value. It serves to provide a leakage path for any current which may flow through the zener before it conducts properly. C_1 serves to remove any transient spikes which may appear on the line. Its value should not be high otherwise the response of the circuit may be slowed down too much. A suitable value is about 10 nF. Although this method of protection may seem rather crude, it does prove to be quick, reliable and effective. In addition to this, it only requires the use of a fuse, zener diode, resistor and thyristor and because of this it has found widespread use.

A Quick and Easy NiCad Charger
With the increase in the number of pieces of portable equipment there is a growing use of NiCads. In fact they are being used more often not only with amateur radio equipment but also in a whole range of battery powered goods. Their obvious advantage is that they can be re-used by recharging them when they run down. A charger, however, is not always available. In cases like these, it is possible to use a standard PSU with a handful of other components to make a simple but effective charger. This can be a particularly attractive proposition because many stations which use NiCads are likely to possess a 12 volt power supply for powering various pieces of base station equipment.

The circuit for the simple charger is shown in Figure 5.22. The resistor must be included because the NiCad cells have a very low internal resistance. It must be chosen to limit the current to the value recommended by the manufacturer so that the cells are not damaged.

The value of the series resistor can be easily calculated if the charging current is known. Take an example where four C cells need to be recharged from the station PSU which gives 13.8 volts. Each cell will have a voltage of 1.2 volts giving a total of 4.8 volts. A typical charging current would be 250 mA. Therefore the resistor needs to drop 13.8–4.8 = 9 volts at a current of 250 mA. From this it is easy to calculate the series resistor needs to be 36 ohms and it will have to be capable of dissipating 2.25 watts. The resultant charger is quite effective and has the advantage of costing only a few pence against a commercially made one which might cost about £10.

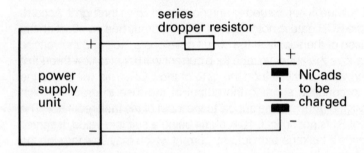

Figure 5.22 The quick and easy NiCad charger

6 TESTING IDEAS

Testing circuits and getting them to work properly is just as important a part of home construction as the actual building. Circuits usually do not work properly first time, and they need a certain amount of persuasion or cajoling before they work. Unfortunately this is one of the more difficult parts of construction as it often requires test equipment as well as a certain amount of know-how about the tricks of the trade.

Additionally, it is useful to be able to maintain some of the station equipment. Although it is often necessary to return some of the more sophisticated up-to-date rigs to a dealer, there is still room for the amateur to have a go, especially if the equipment is second-hand and not in pristine condition.

To carry out maintenance and testing it is necessary to invest in some test equipment. Obviously a test meter is the first essential requirement. This can either be an ordinary analogue multimeter or it may be worth considering one of the cheaper digital versions. What-

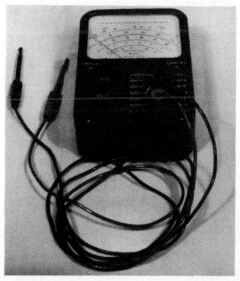

Figure 6.1 A multimeter

ever is chosen it should be robust, as it is the one piece of test equipment which always seems to get dropped.

Another piece of equipment which is almost as invaluable is a GDO. These initials originally stood for grid dip oscillator in the days of valves, but today field effect transistors are used and these are sometimes referred to as FDO's (Fet Dip Oscillators). Whatever their name, they can be used for a wide variety of jobs from detecting the resonance of a circuit to acting as an absorption wavemeter.

Other pieces of test equipment such as oscilloscopes, frequency counters, transistor testers and so forth are also very useful. However, they can be expensive and, unless a lot of servicing or testing is envisaged, they may not earn their keep. Also, it is sometimes possible to improvise and use other methods to test the circuit.

Diode Test

A quick confidence check on semiconductor devices can be very useful from time to time. Obviously the simplest to check is a diode. As it will conduct only in one direction it is quite easy to check using the ohms range on a simple analogue multimeter.

To check that the diode conducts in the forward direction, connect a cathode to the terminal marked positive on the multimeter and the anode to the negative terminal. With the meter set to read ohms the meter should deflect and show a lowish resistance. The actual resistance shown will vary according to the meter, its range and the diode. To give an example an AVO on the x1 ohms range showed a resistance of 30 ohms for a 1N914, but indicated about 1K2 on the x100

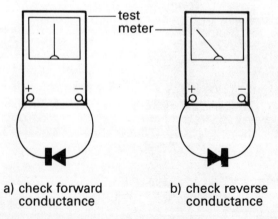

a) check forward
conductance

b) check reverse
conductance

Figure 6.2 Checking a diode with a test meter

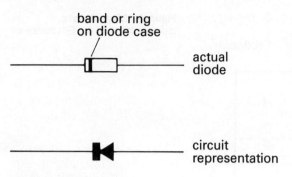

band or ring
on diode case

actual
diode

circuit
representation

Figure 6.3 Diode markings

range, and about 8 K on the x100 range. In each case the meter deflection was greater than about a third of full scale deflection. The main criterion was that the meter did deflect and showed that the diode was conducting.

The reason for this change is simple. It is caused by the fact that the diode is not linear and will have an almost fixed drop of about 0.6 volts across it if it is silicon and 0.3 volts if it is germanium. This will obviously cause the meter to show different readings on different scales.

Once checked in the forward direction it can be tested in the reverse direction by changing the connections round. In this direction the meter should not deflect if it is silicon but, if it is germanium, it may slightly.

This test is useful because it is quick and convenient and gives a good indication of whether the diode is still basically operational. It should be remembered that it will not test such parameters as reverse breakdown and so forth which can cause the diode to fail in certain applications. Because of this, it should be treated only as a confidence check.

A Simple Transistor Check

The diode test can be extended to give a simple transistor test. Again it only provides a good confidence check that the device has not blown, but in spite of this it can be extremely useful

Looking at the construction of a typical NPN transistor in Figure 6.4 it can be seen that it has two PN junctions. This can be represented diagrammatically, as shown in Figure 6.5, showing that it can be tested as two diodes. Whilst the example shows an NPN transistor, the same assumption can be applied to a PNP transistor.

To test the transistor, the diode test must be performed between the base and collector, and base and emitter using an analogue

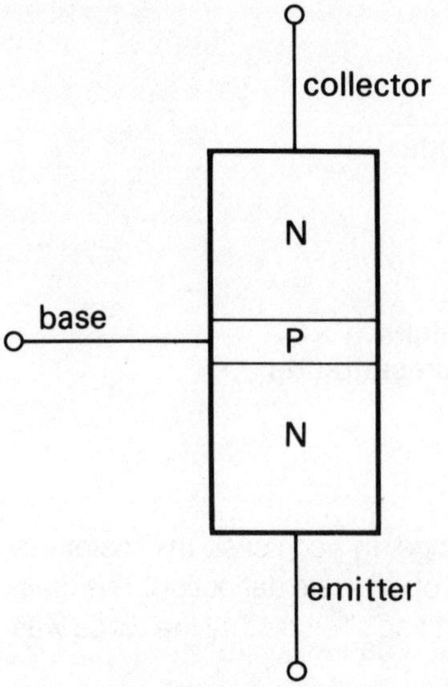

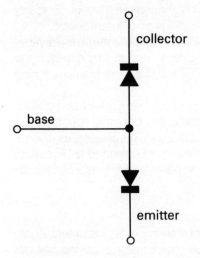

Figure 6.4 Diagrammatic representation of an NPN transistor

collector

N

base   P

N

emitter

Figure 6.5 Representation of an NPN transistor as two back-to-back diodes

collector

base

emitter

meter. Again the reading in the forward direction should be more than about one third of the full scale deflection of the meter, and in the reverse direction it should be negligible. Most small signal silicon devices will give no visible deflection in the reverse direction, but germanium transistors, and silicon power transistors, will show some reading.

One further test is then required. The resistance between collector and emitter must be checked. It should be about the same as the value obtained when measuring each junction in the reverse direction. This final test is required because it is sometimes possible to blow the transistor in such a way that the collector and emitter regions are shorted together, but a diode junction still remains between the base and the collector and emitter. This failure usually occurs if the transistor has passed too much current.

Testing FET's

It is also possible to test some types of FET using this method. JFET's like the popular 2N3819 can be tested, but other forms of FET such as the MOSFET or IGFET cannot. This is because their gates are physically insulated from the channel by a thin oxide layer.

The test hinges around the equivalent circuit of the FET shown in Figure 6.6. The resistors R_1 and R_2 represent the channel resistance and the diode D_1 represents the gate – channel junction.

The first step in the test is to ensure that the channel is intact by using a test meter to measure its resistance. This should be the same in both directions and a reading of around 200 ohms may be obtained. As with many semiconductor parameters this value may vary quite widely from device to device, so it should only be used as a very rough guide.

The next step is to find out whether the junction between the gate and channel is still intact. This is done by measuring the resistance between the gate and either the drain or the source. If this junction is still intact then it should conduct in only one direction.

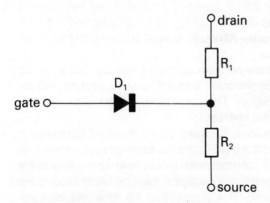

Figure 6.6 Equivalent circuit of a J-FET

This test can also be used to identify whether the device is either an n-channel or p-channel type. If the FET is an n-channel variety then it will conduct when the negative probe on the test meter is connected to the gate and the positive probe is connected to the drain or source. Conversely, for a p-channel FET a low resistance will be indicated when the positive test meter terminal is connected to the gate.

Again this is not a complete test of a FET, but it does give a good indication whether it has blown or not.

Ohms Range Quirk on Multimeters

Most analogue multimeters such as AVO's or the cheaper, more popular meters have a small 'quirk' on the ohms range. Although the terminals marked positive and negative operate as expected on the volts and amps ranges with the positive and negative voltages going to their respective terminals, this is not quite so on the ohms range. Normally, one might expect that the internal batteries used on the ohms range to make the terminal marked positive, positive with respect to the negative one. Unfortunately this is not so because of the internal circuitry of the meter and could give some misleading results. So be warned. On the ohms range, the positive terminal has a negative voltage on it from the internal battery and the negative one has a positive voltage.

A Rough and Ready Signal Generator

Signal generators are often expensive pieces of test gear which few people can afford. In spite of this, they are useful when tracing faults in receivers or other units which require an RF signal. To overcome this, it is often possible to improvise in many ways. One solution can be to run a transmitter into a dummy load and to use the power it radiates. Another possibility is to use the signal radiated from the local oscillator of a nearby radio. Although its level will be quite low, it can be useful in many cases.

Often a spare transistor portable will be a prime candidate for the lower frequencies. Its oscillator will also not be well screened, and will radiate a reasonable signal. The drawbacks are obviously the frequency coverage and the stability.

In order to tune the local oscillator to the required frequency it should be remembered that it will not be the same as that indicated on the receiver dial. The I.F. of most medium and long wave receivers is about 465 KHz, and usually the oscillator is set to operate above the frequency being received. This means that the local oscillator frequency will be about 465 KHz above the signal the radio would receive, ie., about 465 KHz above the frequency indicated on the dials.

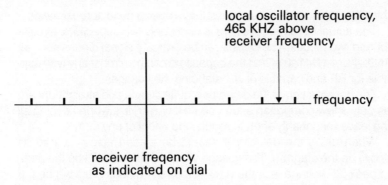

Figure 6.7 Local oscillator of a receiver is usually 465 KHz above indicated receiver frequency

A Simple RF Output Detector

When building or setting up transmitters, it is essential to have some way of monitoring the output. This can be done in several ways. One is to use a VSWR bridge or power meter in the coax lead between the transmitter and its load. Another way might be to monitor the amount of radiated power using an absorption wavemeter, but this would mean radiating a test signal with the possibility of causing interference.

One further way which is quite cheap and is ideal for measuring the output from QRP transmitters, is to use a simple detector circuit as shown in Figure 6.8. Using this, it is possible to measure the peak to peak RF voltage across a load and from this obtain an appropriate value for the output power. Unfortunately, losses and stray effects mean that the detector may not be particularly accurate, especially at

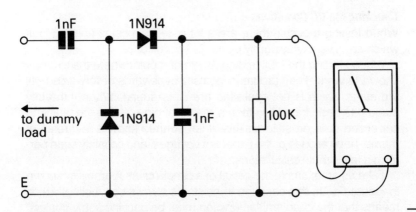

Figure 6.8 Circuit of the simple RF output detector

the higher frequencies, but it will still give quite a good approximation.

The actual circuit is quite simple using only two capacitors, a resistor and two diodes. The diodes can be ordinary signal diodes such as 1N914's or 1N4148's, and the capacitors can be almost any type suitable for RF and capable of withstanding the voltages.

The construction of the detector is quite simple and straightforward as only a few components are used. However, it is worth keeping all the leads reasonably short to reduce the effect of any strays.

When using the detector it should be placed across a load as shown in the diagram. The voltage which it indicates will be the peak to peak RF voltage less the drop across the diodes. This will be 0.6 volt per diode if they are silicon or 0.2 volts per diode in the case of germanium ones. So, for the case of the detector using 1N914 or 1N4148 diodes the actual voltage will be 1.2 volts more than the indicated value. This value must then be converted to an RMS voltage by dividing it by 2.82. Then the output in watts can be calculated by plugging the RMS voltage into the formula $V^2 = RW$. Alternatively if silicon diodes are used together with a 50 ohm load, the whole calculation can be performed by simply plugging the voltage reading obtained on the voltmeter into the formula:

$$W = \frac{(V + 1.2)^2}{141}$$

The value obtained from this should not be treated as exact, especially at higher frequencies. However, it does give a good guide to the output power, and it can be used quite satisfactorily to tune for maximum output.

Checking the DC Conditions

When testing a circuit, there are a lot of tell-tale signs about a fault which are given away purely by the DC conditions.

When starting the fault finding, begin at a point where the circuit is known to work. Then progress logically backwards or forwards until the stage which is not operating has been found. When the faulty stage has been found it is then a matter of pin-pointing the fault and correcting it. In order to do this, it is useful to know a few rules of thumb, have an idea of the expected voltages, and possibly even perform a few simple calculations.

Take as an example the circuit of a simple class A amplifier shown in Figure 6.9. In this case the transistor is always conducting, which means that the base emitter junction must be passing some current. When this happens, it is found that the base potential will be 0.6 volts

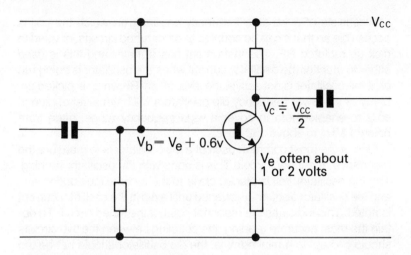

Figure 6.9 A typical class A amplifier stage

higher than the emitter if it is a silicon transistor, or 0.2 volts if it is germanium. If it is much less or much more, then it should be investigated. In fact any silicon, be it switching or analogue, will have the base sitting at 0.6 volts above the emitter when there is current flowing between the collector and emitter.

Another check worth performing is to look at the collector voltage to find out what it should be. In the case of the analogue circuit it should normally sit at about half the supply voltage. This is so that, when an audio signal is applied, the collector voltage can vary equally either side of its quiescent value without distorting.

In the case of the switching circuit, it is worth ensuring that the transistor is fully on or fully off. If this is the case, then the collector voltage should be virtually zero, or the supply voltage. This test should be done with a static input level to the circuit. If a switching waveform is applied to the circuit then the meter will only respond to the average and this may be misleading. Simple measurements like these can detect a large number of faults. Obviously they will not be able to detect all of them, but they do give a good starting point.

Making the Most of a Dip Oscillator

Aside from a multimeter, a dip oscillator (GDO or FDO) must be one of the most useful and versatile pieces of test equipment available to the radio amateur. It can be used for a wide variety of jobs, from determining the resonant frequency of a tuned circuit to use as an absorption wavemeter or signal monitor.

It consists of a variable frequency oscillator on which the coil is accessible so that it can be coupled to other tuned circuits, or used to pick up radiated RF. The instrument has a meter and this is used either to monitor the oscillator current when the oscillator is being run or, if the oscillator is not active, the level of any RF which is picked up. Many of the lower frequency dip oscillators will have a set of plug-in coils to enable them to cover a wide frequency range, often from below 1 MHz to above 150 MHz.

One of the most common uses of a dip oscillator is to measure the resonance of a tuned circuit. This is done with the oscillator running. The dip oscillator coil is placed close to the tuned circuit under test, and the oscillator frequency is varied until a dip in the oscillator current is noted. This indicates the resonant point of the tuned circuit. To obtain the most accurate reading, the coupling between the two circuits should be kept to a minimum, ie. the dip oscillator should not be too close to the circuit under test. Ideally, the dip oscillator should be moved away until a small but clearly visible dip is obtained. Another point worth noting is that, if there are several coils in any area, one should make sure that the dip is being caused by the right coil. It is possible to check this by moving the dip oscillator around and checking that the dip is strongest in the vicinity of the coil being measured.

As it is possible to measure the resonant frequency of a tuned circuit, it is possible to use this information to determine the value of an unknown inductor or capacitor. If an unknown inductor is to be measured, then a capacitor of a known value should be placed in parallel with it. The resonant frequency should be determined, and then the value of the inductor can be calculated from the formula

$$f = \frac{1}{2\pi \sqrt{LC}}$$

In this formula f is the resonant frequency in Hertz, L the inductance in Henries and C the capacitance in Farads.

If the value of a capacitor is to be determined, the same method can be used, the only difference being that the capacitance is unknown instead of the inductance. As the values of inductors are not always marked, this may mean that the value of the inductor has to be determined first.

Another valuable use for a dip oscillator is as a signal generator. Place it near the aerial socket of the receiver which is being tested and the signal level can then be varied by altering the distance between the oscillator and the receiver.

The dip oscillator can also be used as an absorption wavemeter. When used in this mode, the oscillator is not active and the meter is used to indicate the detected RF. The dip meter is placed close to the source of RF and tuned until it indicates a peak. It is particularly useful here to check that the signal is in the right band and that the correct harmonic or mix product has been selected. As it is fairly insensitive, a dip meter would normally be used only in conjunction with transmitters to check that they are approximately on the correct frequency, and that they are not radiating large amounts of spurious signals.

Even though it does have some limitations a dip oscillator, GDO, FDO or call it what you will, can be used in a large number of ways and is a useful addition to any radio station.

Figure 6.10 A dip oscillator

APPENDIX 1

SUMMARY OF BASIC FORMULAE

Ohms Law
The relationship of the voltage V across a resistor, the current I through it and its resistance R can be expressed as

$$V = IR$$

Power Calculations
The power W dissipated in a resistor can be expressed as

$$W = VI$$

By using the formula for Ohms Law it can also be written as

$$W = I^2R$$

or

$$W = \frac{V^2 R}{R}$$

Conductance
The conductance G of a resistor is the reciprocal of resistance and its units are mhos

$$G = \frac{I}{R}$$

Ohms Law becomes

$$V = \frac{I}{G}$$

Combinations of Resistors
In series

$$R_{TOTAL} = R_1 + R_2 + R_3 \ldots$$

In parallel

$$\frac{1}{R_{TOTAL}} = \frac{1}{R_1} + \frac{1}{R_2} + \frac{1}{R_3} \ldots$$

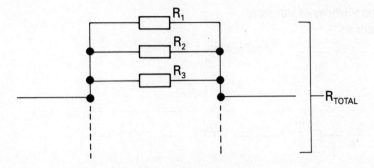

If there are only two resistors this simplifies to

$$R_{TOTAL} = \frac{R_1 R_2}{= R_1 + R_2}$$

Combinations of Capacitors

In parallel

$$C_{TOTAL} = C_1 + C_2 + C_3 \ldots$$

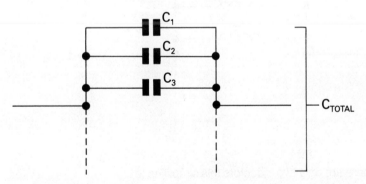

In series

$$\frac{1}{C_{TOTAL}} = \frac{1}{C_1} + \frac{1}{C_2} + \frac{1}{C_3} \ldots$$

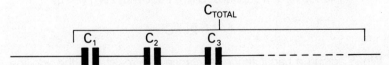

If there are only two capacitors this simplifies to

$$C_{TOTAL} = \frac{C_1 \, C_2}{C_1 + C_2}$$

Combinations of Inductors

In series

$$L_{TOTAL} = L_1 + L_2 + L_3 \ldots$$

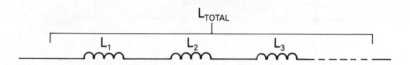

In parallel

$$\frac{1}{L_{TOTAL}} = \frac{1}{L_1} + \frac{1}{L_2} + \frac{1}{L_3} \ldots$$

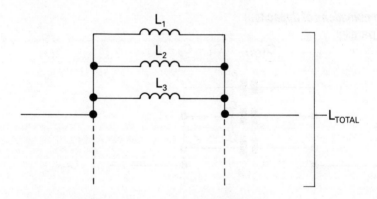

If there are only two inductors this simplifies to

$$L_{TOTAL} = \frac{L_1 \, L_2}{L_1 + L_2}$$

Reactance and Impedance

The reactance X_c of a capacitor C at a frequency f is given by

$$X_c = \frac{1}{2\pi f C}$$

When adding a capacitance reactance to a resistance the addition has to be done vectorially. The total impedance Z_{TOTAL} has to be calculated as below

$$Z_{TOTAL} = \sqrt{X_c^2 + R^2}$$

The reactance X_L of an inductor L at a frequency f is given by

$$X_L = 2\pi fL$$

When adding an inductive reactance to a resistance this also has to be done vectorially

$$Z_{TOTAL} = \sqrt{X_L^2 + R^2}$$

Admittance
The admittance Y of a circuit is the reciprocal of its impedance

$$Y = \frac{1}{Z}$$

Resonant Circuits
Resonance occurs in a circuit when the capacitive and inductive reactances are equal. The frequency f can be determined from the formula

$$f = \frac{1}{2\pi\sqrt{LC}}$$

The Q or quality factor of a resonant circuit is particularly important when determining bandwidths and losses in the circuit. There are two useful formulae associated with definitions of Q

$$Q = \frac{2\pi L}{R}$$

$$Q = \frac{\triangle f}{f_c}$$

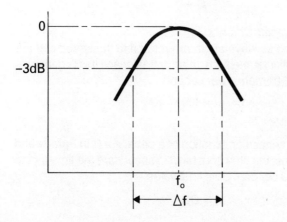

Power Level Comparisons

Figures of gain or loss in circuits are usually expressed in terms of decibels.

$$N \, dB = 10 \log_{10} \frac{P_2}{P_1}$$

P_1 and P_2 are the two power levels being compared.

If these figures are expressed in terms of voltages V_1 and V_2 the formula changes

$$N \, dB = 20 \, Log_{10} \frac{V_2}{V_1}$$

Similarly for two currents I_1 and I_2

$$N \, dB = 20 \, Log_{10} \frac{I_2}{I_1}$$

If the impedances of the two points being compared are different this has also to be taken into consideration

$$N \, dB = 20 \log_{10} \frac{V_2}{V_1} + 10 \log_{10} \frac{Z_1}{Z_2}$$

Z_1 and Z_2 are the impedances at the points where V_1 and V_2 were measured.

VSWR

When using any system which transfers RF power from one point to another the VSWR is of importance. It can be calculated from a knowledge of the reflection coefficient ρ:-

$$VSWR = S = \frac{1 + \rho}{1 - \rho}$$

Conversely

$$\rho = \frac{S - 1}{S + 1}$$

Frequency to Wavelength Conversion

The frequency f and its wavelength are related to the speed c of the radio wave in a particular medium. In air or free space this is generally taken to be 300×10^6 metres per second.

$$\lambda f = c = 300 \times 10^6$$

Time Constant

In a circuit using a series combination of a capacitor C in Farads and a resistor R in Ohms, the time constant t in seconds is the time for the voltage to rise to 63 per cent of its final value

$$t = CR$$

Similarly for a circuit containing an inductor L in Henries in series with a resistor R in Ohms, the time constant t in seconds is the time for the current to rise to 63 per cent of its final value

$$t = \frac{L}{R}$$

Transformer Ratios

The impedance of the primary circuit Zp of a transformer and the impedance of the secondary circuit Zs are related to the number of turns on the primary Np and the number of turns on the secondary Ns

$$\frac{Np}{Ns} = \sqrt{\frac{Zp}{Zs}}$$

APPENDIX 2

Conversion and Reference Tables

DECIBEL TO POWER, VOLTAGE OR CURRENT RATIO CONVERSION TABLE

dB	Power Ratio	Voltage or Current Ratio
0.1	1.023	1.012
0.2	1.047	1.023
0.3	1.072	1.035
0.4	1.096	1.047
0.5	1.122	1.059
0.6	1.148	1.072
0.7	1.175	1.084
0.8	1.202	1.096
0.9	1.230	1.109
1.0	1.259	1.122
2.0	1.585	1.259
3.0	1.995	1.413
4.0	2.812	1.585
5.0	3.162	1.778
6.0	3.981	1.995
7.0	5.012	2.339
8.0	6.310	2.512
9.0	7.943	2.818
10.0	10.000	3.162
20	10^2	10.000
30	10^3	31.623
40	10^4	100.00
50	10^5	316.22
60	10^6	1000.0
70	10^7	3162.2
80	10^8	10000
90	10^9	31622

dBm – dBW WATTS CONVERSION TABLE

dBm	dBW	Watts	Terminology
+100	+70	10000000	10 Megawatts
+90	+60	1000000	1 Megawatt
+80	+50	100000	100 Kilowatts
+70	+40	10000	10 Kilowatts
+60	+30	1000	1 Kilowatt
+50	+20	100	100 Watts
+40	+10	10	10 Watts
+30	0	1	1 Watt
+20	−10	0.1	100 Milliwatts
+10	−20	0.01	10 Milliwatts
0	−30	0.001	1 Milliwatt
−10	−40	0.0001	100 Microwatts
−20	−50	0.00001	10 Microwatts
−30	−60	0.000001	1 Microwatt
−40	−70	0.0000001	100 Nanowatts
−50	−80	0.00000001	10 Nanowatts
−60	−90	0.000000001	1 Nanowatt

WATTS TO VOLTS CONVERSION TABLE (FOR 50 OHM TERMINATED SYSTEM)

Watts	Volts (RMS)	Volts (Peak)	Volts (Peak to Peak)
0.1	2.24	3.16	6.32
0.2	3.16	4.47	8.94
0.3	3.87	5.48	11.0
0.4	4.47	6.32	12.7
0.5	5.00	7.07	14.1
0.6	5.48	7.45	15.5
0.7	5.92	8.37	16.7
0.8	6.32	8.94	17.9
0.9	6.71	9.49	19.0
1.0	7.07	10.0	20.0
2.0	10.0	14.1	28.3
3.0	12.2	17.3	34.6
4.0	14.1	20.0	40.0
5.0	15.8	22.4	44.7
6.0	17.3	24.5	49.0
7.0	18.7	26.5	52.9
8.0	20.0	28.3	56.6
9.0	21.2	30.0	60.0
10	22.4	31.6	63.2
20	31.6	44.7	89.4
30	38.7	54.8	110
40	44.7	63.2	127
50	50.0	70.7	141
60	54.8	74.5	155
70	59.2	83.7	167
80	63.2	89.4	179
90	67.1	94.9	190
100	70.7	100	200

VSWR TO REFLECTION COEFFICIENT CONVERSION TABLE

VSWR	Reflection Coefficient ρ
1.1	0.05
1.2	0.09
1.3	0.13
1.4	0.17
1.5	0.20
1.6	0.23
1.7	0.26
1.8	0.29
1.9	0.31
2.0	0.33
2.5	0.43
3.0	0.50
3.5	0.56
4.0	0.60
4.5	0.64
5.0	0.67
6.0	0.71
7.0	0.75
8.0	0.78
9.0	0.80
10.0	0.82
15.0	0.87
20.0	0.90

FREQUENCY TO WAVELENGTH CONVERSION TABLE

Frequency (MHz)	Wavelength
0.2	1500 m
0.3	1000 m
0.5	600 m
0.7	400 m
1	300 m
2	150 m
3	100 m
5	60 m
7	40 m
10	30 m
20	15 m
30	10 m
50	6 m
70	4 m
100	3 m
200	1.5 m
300	1 m
500	60 cm
700	40 cm
1000	30 cm
2000	15 cm
3000	10 cm
5000	6 cm
7000	4 cm
10000	3 cm

IMPERIAL TO METRIC LENGTH CONVERSION TABLE

Fraction of Inch	Inches	Millimetres
1/32	0.031	0.794
1/16	0.063	1.588
3/32	0.094	2.381
1/8	0.125	3.175
5/32	0.156	3.969
3/16	0.188	4.763
7/32	0.219	5.556
1/4	0.250	6.350
9/32	0.281	7.144
5/16	0.313	7.938
11/32	0.344	8.731
3/8	0.375	9.525
13/32	0.406	10.32
7/16	0.438	11.11
15/32	0.469	11.91
1/2	0.500	12.70
17/32	0.531	13.49
9/16	0.563	14.29
19/32	0.594	15.08
5/8	0.625	15.88
21/32	0.656	16.67
11/16	0.688	17.46
23/32	0.719	18.26
3/4	0.750	19.05
25/32	0.781	19.84
13/16	0.813	20.64
27/32	0.844	21.43
7/8	0.875	22.23
29/32	0.906	23.02
15/16	0.938	23.81
31/32	0.969	24.61
1	1.000	25.40

EFFECT OF V.S.W.R. ON TRANSMITTED POWER

V.S.W.R.	Proportion of Power Transmitted (%)
1 : 1.0	100
1 : 1.1	99.8
1 : 1.2	99.2
1 : 1.5	96.0
1 : 2.0	89
1 : 3.0	75
1 : 4.0	64
1 : 5.0	55
1 : 10.0	32
1 : 15	24
1 : 20	19

TABLE OF INDUCTOR REACTANCE (IN OHMS)

Inductance \ Frequency	1.8MHz	3.5MHz	7MHz	14MHz	21MHz	28MHz	50MHz	70MHz	144MHz
100nH	1.1	2.2	4.4	8.8	13.2	17.6	31.4	43.6	90.5
470nH	5.3	10.3	20.7	41.3	62.0	82.7	147.7	207	425
1μH	11.3	22.0	44.0	88.0	132	176	314	436	905
4.7μH	53.2	103	207	413	620	827	1477	2067	4252
10μH	113	220	440	880	1319	1759	3142	4398	9048
47μH	532	1034	2067	4134	6202	8269	14770	20670	42520
100μH	1131	2199	4398	8796	13190	17590	31420	43980	90480
470μH	5316	10340	20670	41340	62020	82690	147.7K	206.7K	425.2K
1000μH	11310	21990	43980	87960	131.9K	175.9K	314.2K	439.8K	904.8K

NB: At high frequencies and large inductances the inductor self capacitance will predominate

TABLE OF CAPACITOR REACTANCE (IN OHMS)

Capacitance \ Frequency	1.8MHz	3.5MHz	7MHz	14MHz	21MHz	28MHz	50MHz	70MHz	144MHz
1pF	88420	45470	22740	11370	7579	5684	3183	2274	1105
4.7pF	18810	9675	4838	2418	1613	1209	677	484	235
10pF	8842	4547	2274	1137	758	568	318	227	111
47pF	1881	968	484	242	161	121	67.7	48.4	23.5
100pF	884	455	227	114	75.8	56.8	31.8	22.7	11.1
470pF	188	96.8	48.4	24.2	16.1	12.1	6.8	4.8	2.4
1nF	88.4	45.5	22.7	11.4	7.6	5.7	3.2	2.3	1.1
4.7nF	18.8	9.7	4.8	2.4	1.6	1.2	0.7	0.5	0.2
10nF	8.8	4.6	2.3	1.1	0.8	0.6	0.3	0.2	0.1
47nF	1.8	0.9	0.5	0.2	0.2	0.1	0.1	—	—
100nF	0.9	0.5	0.2	0.1	0.1	0.1	—	—	—

APPENDIX 3

General Information

MULTIPLES AND SUB-MULTIPLES

Factor	Prefix	Abbreviation
10^{12}	Tera	T
10^{9}	Giga	G
10^{6}	Mega	M
10^{3}	Kilo	k
10^{2}	Hecto	h
10	Deka	dk
10^{-1}	Deci	d
10^{-2}	Centi	c
10^{-3}	Milli	m
10^{-6}	Micro	μ
10^{-9}	Nano	n
10^{-12}	Pico	p

UNITS

Quantity	Unit Name	Unit Symbol
Capacitance	Farad	F
Charge	Coulomb	As (Ampere seconds)
Current	Ampere	A
Electric Field	Volt/metre	V/m
Frequency	Hertz	Hz
Inductance	Henry	H
Length	Metre	m
Magnetic Field	Ampere/metre	A/m
Potential Difference	Volt	V
Power	Watt	W
Resistance	Ohm	Ω
Time	Second	s

RADIO FREQUENCY CLASSIFICATIONS

Frequency Band	Designation	Abbreviation
3–30 kHz	Very Low Frequency	VLF
30–300 kHz	Low Frequency	LF
300–3000 kHz	Medium Frequency	MF
3–30 MHz	High Frequency	HF
30–300 MHz	Very High Frequency	VHF
300–3000 MHz	Ultra High Frequency	UHF
3–30 GHz	Super High Frequency	SHF
30–300 GHz	Extremely High Frequency	EHF

U.K. AMATEUR FREQUENCY ALLOCATIONS

1.81	–	2.00 MHz
3.50	–	3.80 MHz
7.00	–	7.10 MHz
10.10	–	10.15 MHz
14.00	–	14.35 MHz
18.068	–	18.168 MHz
21.00	–	21.45 MHz
24.89	–	24.99 MHz
28.00	–	29.70 MHz
50.00	–	52.00 MHz
70.025	–	70.50 MHz
144.00	–	146.00 MHz
430.00	–	440.00 MHz
1240.0	–	1325.0 MHz
2310.0	–	2450.0 MHz
3400.0	–	3475.0 MHz
5650.0	–	5680.0 MHz
5755.0	–	5765.0 MHz
5820.0	–	5850.0 MHz
10000	–	10500 MHz
24000	–	24250 MHz
47000	–	47200 MHz
75500	–	76000 MHz
142000	–	144000 MHz
248000	–	250000 MHz

Class B licencees may only use frequencies above 30 MHz.
Class A licencees have access to all frequencies.

LONG, MEDIUM AND SHORTWAVE BANDS

Long Wave	0.0150	–	0.285 MHz
Medium Wave	0.5265	–	1.6065 MHz
120 Metres	2.300	–	2.498 MHz
90 Metres	3.200	–	3.400 MHz
75 Metres	3.950	–	4.000 MHz
60 Metres	4.750	–	5.060 MHz
49 Metres	5.950	–	6.200 MHz
41 Metres	7.100	–	7.300 MHz
31 Metres	9.500	–	9.900 MHz
25 Metres	11.650	–	12.050 MHz
22 Metres	13.600	–	13.800 MHz
19 Metres	15.100	–	15.600 MHz
16 Metres	17.550	–	17.900 MHz
13 Metres	21.450	–	21.850 MHz
11 Metres	25.670	–	26.100 MHz

UK BROADCAST BANDS ABOVE 30 MHz

Band No.	Frequency Range MHz	Channel Nos.	Uses
Band I	41–68	1–5	405 line TV*
Band II	88–108	—	FM Radio
Band III	174–216	6–13	405 line TV*
Band IV	470–582	21–34	625 line TV
Band V	614–854	39–68	625 line TV
Band VI	11700–12500	1–40	Satellite TV

*405 line TV transmissions have now been discontinued in the UK. These frequencies are now used for other services.

UK CITIZEN'S BAND CHANNEL FREQUENCIES

27 MHz Band

Channel No.	Frequency MHz	Channel No.	Frequency MHz
1	27.60125	21	27.80125
2	27.61125	22	27.81125
3	27.62125	23	27.82125
4	27.63125	24	27.83125
5	27.64125	25	27.84125
6	27.65125	26	27.85125
7	27.66125	27	27.86125
8	27.67125	28	27.87125
9	27.68125	29	27.88125
10	27.69125	30	27.89125
11	27.70125	31	27.90125
12	27.71125	32	27.91125
13	27.72125	33	27.92125
14	27.73125	34	27.93125
15	27.74125	35	27.94125
16	27.75125	36	27.95125
17	27.76125	37	27.96125
18	27.77125	38	27.97125
19	27.78125	39	27.98125
20	27.79125	40	27.99125

UK CITIZEN'S BAND CHANNEL FREQUENCIES

934 MHz Band

Channel No.	Frequency MHz	Channel No.	Frequency MHz
1	934.025	11	934.525
2	934.075	12	934.575
3	934.125	13	934.625
4	934.175	14	934.675
5	934.225	15	934.725
6	934.275	16	934.775
7	934.325	17	934.825
8	934.375	18	934.875
9	934.425	19	934.925
10	934.475	20	934.975

A	·—		N	—·
B	—···		O	———
C	—·—·		P	·——·
D	—··		Q	——·—
E	·		R	·—·
F	··—·		S	···
G	——·		T	—
H	····		U	··—
I	··		V	···—
J	·———		W	·——
K	—·—		X	—··—
L	·—··		Y	—·——
M	——		Z	——··

Number	Code		Number	Code
1	·————		6	—····
2	··———		7	——···
3	···——		8	———··
4	····—		9	————·
5	·····		0	—————

Full stop	·—·—·—		Comma	——··——
?	··——··		=	—···—
Wait	·—···		Mistake	········
Stroke (/)	—··—·			

Start of Work (C̅T̅)	—·—·—		End of Message (A̅R̅)	·—·—·
Invitation to Transmit (K)	—·—		Invitation for a	
End of Work (V̅A̅)	···—·—		particular station to transmit (K̅N̅)	—·——·

PHONETIC ALPHABET

A	Alpha		N	November
B	Bravo		O	Oscar
C	Charlie		P	Papa
D	Delta		Q	Quebec
E	Echo		R	Romeo
F	Foxtrot		S	Sierra
G	Golf		T	Tango
H	Hotel		U	Uniform
I	India		V	Victor
J	Juliet		W	Whisky
K	Kilo		X	X-Ray
L	Lima		Y	Yankee
M	Mike		Z	Zulu

PREFERRED VALUES FOR PASSIVE COMPONENTS

E6 Series	1.0	1.5	2.2	3.3	4.7	6.8
E12 Series	1.0	1.2	1.5	1.8	2.2	2.7
	3.3	3.9	4.7	5.6	6.8	8.2
E24 Series	1.0	1.1	1.2	1.3	1.5	1.6
	1.8	2.0	2.2	2.4	2.7	3.0
	3.3	3.6	3.9	4.3	4.7	5.1
	5.6	6.2	6.8	7.5	8.2	9.1
E48 Series	1.00	1.05	1.10	1.15	1.21	1.27
	1.33	1.40	1.47	1.54	1.62	1.69
	1.78	1.87	1.96	2.05	2.15	2.26
	2.37	2.49	2.61	2.74	2.87	3.01
	3.16	3.32	3.48	3.65	3.83	4.02
	4.22	4.42	4.64	4.87	5.11	5.36
	5.62	5.90	6.19	6.49	6.81	7.15
	7.50	7.87	8.25	8.66	9.09	9.53

PREFERRED VALUES FOR PASSIVE COMPONENTS

E96 Series	1.00	1.02	1.05	1.07	1.10	1.13
	1.15	1.18	1.21	1.24	1.27	1.30
	1.33	1.37	1.40	1.43	1.47	1.50
	1.54	1.58	1.62	1.65	1.69	1.74
	1.78	1.82	1.87	1.91	1.96	2.00
	2.05	2.10	2.15	2.21	2.26	2.32
	2.37	2.43	2.49	2.55	2.61	2.67
	2.74	2.80	2.87	2.94	3.01	3.09
	3.16	3.24	3.32	3.40	3.48	3.57
	3.65	3.74	3.83	3.92	4.00	4.12
	4.22	4.32	4.42	4.53	4.64	4.75
	4.87	4.99	5.11	5.23	5.36	5.49
	5.62	5.76	5.90	6.04	6.19	6.34
	6.49	6.65	6.81	6.98	7.15	7.32
	7.50	7.68	7.87	8.06	8.25	8.45
	8.66	8.87	9.09	9.31	9.53	9.76

Connector Type	Impedance (Ohms)	Description
Belling Lee	75	An economy connector used almost exclusively for television equipment. Its main advantage is the low cost.
BNC	50 (75 ohm version available)	A bayonet fitting connector suitable for use up to 10 GHz. It is widely used on professional equipment, although it is not particularly common on commercially produced amateur equipment.
Miniature BNC	50	A miniature version of the ordinary BNC connector.
N	50 (75 ohm version available)	A screw fitting connector suitable for use up to frequencies of 10 GHz and sometimes above. It is often used in situations where good RF performance is essential, and it is often used in high power applications above 300 MHz.
SMA	50	A high quality miniature screw fitting connector for use up to 18 GHz. It is found in many professional microwave applications where miniature flexible or semi-rigid cables are used.
SMB	50	A high quality subminiature snap-on type connector suitable for use up to 18 GHz. It is very convenient to use in similar applications to the SMA or SMC connectors, but it is not as robust.
SMC	50	A sub-miniature connector similar in many respects to the SMA connector. However, it is smaller.
TNC	50	A screw fitting connector very similar in size and characteristics to the BNC connector.
UHF (PL259/SO239)	50	A non-constant impedance connector which is widely used on amateur equipment up to 146 MHz. At frequencies above 200 MHz it introduces a high VSWR.

COMMON SYMBOLS AND CODINGS USED WITH
TRANSISTOR SPECIFICATIONS

f_T	The minimum common emitter gain bandwidth product, i.e. the frequency at which the gain falls to unity in the common emitter configuration.
h_{FE}	DC forward transfer characteristic of a transistor in common emitter configuration. Approximately its DC forward current gain.
h_{fe}	AC forward transfer characteristic of a transistor in common emitter configuration. Approximately its AC forward current gain.
$I_{c\,max}$	Maximum rated collector current.
$P_{TOT\,max}$	Maximum rated device power dissipation.
$T_{J\,max}$	Maximum junction temperature.
V_{cbo}	Collector-base voltage rating with emitter open circuit.
V_{ceo}	Collector–emitter voltage rating with base open circuit.
$V_{ceo\,sat}$	Saturated collector–emitter voltage, i.e. voltage across the transistor when it is turned hard on.
V_{ebo}	Emitter-base voltage rating with collector open circuit.

LOGIC TRUTH TABLES

Inputs		Outputs				
A	B	AND	NAND	OR	NOR	EXCLUSIVE OR
0	0	0	1	0	1	0
1	0	0	1	1	0	1
0	1	0	1	1	0	1
1	1	1	0	1	0	0

FAN-OUT CAPABILITIES BETWEEN DIFFERENT LOGIC FAMILIES

TO	74	74H	74S	74LS	74ALS	74HC
FROM						
74	10	8	8	20	20	*
74H	12	10	10	25	25	*
74S	12	10	10	50	50	*
74LS	5	4	4	20	20	*
74ALS	5	4	4	20	20	*
74HC	2	2	2	12	20	*

*in excess of 1000

ASCII CODES

| ASCII CODE | | KEYBOARD | |
HEX	DEC	Keypress	Character
00	00	CTRL@	NUL
01	01	CTRL-A	SOH
02	02	CTRL-B	STX
03	03	CRTL-C	ETX
04	04	CTRL-D	EOT
05	05	CTRL-E	ENQ
06	06	CTRL-F	ACK
07	07	CTRL-G	BEL
08	08	CTRL-H	BS
09	09	CTRL-I	HT
0A	10	CTRL-J	LF
0B	11	CTRL-K	VT
0C	12	CTRL-L	FF
0D	13	CTRL-M	CR
0E	14	CTRL-N	SO
0F	15	CTRL-O	SI
10	16	CTRL-P	DLE
11	17	CTRL-Q	DC1(Xon)
12	18	CTRL-R	DC2
13	19	CTRL-S	DC3(Xoff)
14	20	CTRL-T	DC4
15	21	CTRL-U	NAK
16	22	CTRL-V	SYN
17	23	CTRL-W	ETB
18	24	CTRL-X	CAN
19	25	CTRL-Y	EM
1A	26	CTRL-Z	SUB
1B	27	CTRL-[	ESC
1C	28	CTRL-\	FS
1D	29	CTRL-]	GS
1E	30	CTRL-∧	RS
1F	31	CTRL-__	US

ASCII CODES

| ASCII CODE | | KEYBOARD | |
HEX	DEC	Keypress	Character
20	32	Space	SP
21	33	!	!
22	34	"	"
23	35	£	£
24	36	$	$
25	37	%	%
26	38	&	&
27	39	'	'
28	40	(	(
29	41	)	)
2A	42	*	*
2B	43	+	+
2C	44	,	,
2D	45	−	−
2E	46	.	.
2F	47	/	/
30	48	0	0
31	49	1	1
32	50	2	2
33	51	3	3
34	52	4	4
35	53	5	5
36	54	6	6
37	55	7	7
38	56	8	8
39	57	9	9
3A	58	:	:
3B	59	;	;
3C	60	<	<
3D	61	=	=
3E	62	>	>
3F	63	?	?

ASCII CODES

ASCII CODE		KEYBOARD	
HEX	DEC	Keypress	Character
40	64	@	@
41	65	A	A
42	66	B	B
43	67	C	C
44	68	D	D
45	&)	E	E
46	70	F	F
47	71	G	G
48	72	H	H
49	73	I	I
4A	74	J	J
4B	75	K	K
4C	76	L	L
4D	77	M	M
4E	78	N	N
4F	79	O	O
50	80	P	P
51	81	Q	Q
52	82	R	R
53	83	S	S
54	84	T	T
55	85	U	U
56	86	V	V
57	87	W	W
58	88	X	X
59	89	Y	Y
5A	90	Z	Z
5B	91	[	[
5C	92	\	\
5D	93	]	]
5E	94	∧	∧
5F	95	—	—

ASCII CODES

ASCII CODE		KEYBOARD	
HEX	DEC	Keypress	Character
60	96	`	`
61	97	a	a
62	98	b	b
63	99	c	c
64	100	d	d
65	101	e	e
66	102	f	f
67	103	g	g
68	104	h	h
69	105	i	i
6A	106	j	j
6B	107	k	k
6C	108	l	l
6D	109	m	m
6E	110	n	n
6F	111	o	o
70	112	p	p
71	113	q	q
72	114	r	r
73	115	s	s
74	116	t	t
75	117	u	u
76	118	v	v
77	119	w	w
78	120	x	x
79	121	y	y
7A	122	z	z
7B	123	{	{
7C	124				
7D	125	}	}
7E	126	—	—
7F	127	Del	Del

NUL	Null (blank)
SOH	Start of header
STX	Start of text
ETX	End of text
EOT	End of transmission
ENQ	Enquiry
ACK	Acknowledge
BEL	Audible bell
BS	Backspace
HT	Horizontal tab
LF	Line feed
VT	Vertical tab
FF	Form feed
CR	Carriage return
SO	Shift out
SI	Shift in
DLE	Data link escape
DC1	Device control 1
DC2	Device control 2
DC3	Device control 3
DC4	Device control 4
NAK	Negative acknowledge
SYN	Synchronous idle
ETB	End of transmission block
CAN	Cancel
EM	End of medium
SUB	Substitute
ESC	Escape
FS	File separator
GS	Group separator
RS	Record separator
US	Unit separator
DEL	Delete

V24/RS232 CONNECTIONS

Connector Pin No.	Circuit Name	Abbreviation
1	Protective Earth	
2	Transmit Data	TXD
3	Receive Data	RXD
4	Request to Send	RTS
5	Clear to Send	CTS
6	Data Set Ready	DSR
7	Signal Ground	
8	Data Carrier Detected	DCD
9		
10		
11	Select Transmit Frequency	
12	Secondary Channel Signal Detect	SDCD
13	Secondary Channel Clear to Send	SCTS
14	Secondary Channel Transmit Data	STXD
15	Transmit Signal Element Timing	TX CLK
16	Secondary Channel Receive Data	SRXD
17	Receive Signal Element Timing	RCV CLK
18	Local Loopback	LL
19	Secondary Channel Request to Send	SRTS
20	Data Terminal Ready, Connect Data Set to Line	DTR, CDSTL
21	Remote Loopback	RL
22	Ringing Indicator, Calling Indicator	RI, CI
23	Data Signalling Rate Selector	
24	External Clock	EXT CLK
25	Test Indicator	

The connector most commonly used is a 25 way D Type.

INDEX

Take out the uncertainty
... take out a subscription